THE SECRET POWER OF THERAPEUTIC PEPTIDES

HOW TO USE INNOVATIVE SCIENCE TO BUILD MUSCLE FASTER, HEAL FROM INJURY, AND BOOST YOUR FOCUS

LUMINARY LIFE

© **Copyright—Luminary Life 2023—All rights reserved.**

The content contained within this book may not be reproduced, duplicated, or transmitted without direct written permission from the author or the publisher.

Under no circumstances will any blame or legal responsibility be held against the publisher or author for any damages, reparation, or monetary loss due to the information contained within this book, either directly or indirectly. You are responsible for your own choices, actions, and results.

Legal Notice:

This book is copyright-protected. This book is only for personal use. You cannot amend, distribute, sell, use, quote, or paraphrase any part of the content within this book without the author's or publisher's consent.

Disclaimer Notice:

Please note the information contained within this book is intended for informational and educational purposes only. While all efforts have been made to present accurate and current information, no guarantees are made regarding its completeness or accuracy. Readers should be aware that the author is not providing medical or professional advice. Please consult a licensed healthcare professional before acting on any information or suggestions in this book.

By reading this book, the reader acknowledges that the author is not liable for any direct, indirect, incidental, or consequential damages or losses arising from the use of the information contained herein, including any errors, omissions, or inaccuracies.

FREE GIFT FOR OUR READERS

Quick Guide: Your 7-Day Immediate Muscle Recovery Peptide Boost Plan

Thank you for picking up a copy of *The Secret Power of Therapeutic Peptides: How to Use Innovative Science to Build Muscle Faster, Heal from Injury, and Boost Your Focus*. As a token of our appreciation, we're giving you access to our guide focused on optimizing muscle recovery using peptides in 7 days.

What will you find inside this Quick Guide?

- Proven strategies to optimize peptide efficiency within a week.
- Daily routines and complementary nutritional tips.
- Key metrics to track your progress.

How to access your copy:

- Scan the below QR code (ebook and print version)
- Click on the below image (ebook version only)

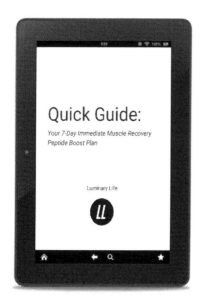

SCAN ME

CONTENTS

Introduction   11

1. UNDERSTANDING PEPTIDES AND
   THEIR ROLE IN THE BODY   19
   Basics of Peptides   20
   Dive into Peptide Biochemistry   25
   Classes of Peptides   30
   Therapeutic Peptides: The Next Wave in
   Health and Fitness   35
   Harnessing the Power of Peptides   37

2. MUSCLES, RECOVERY, AND THE
   MAGIC OF PEPTIDES   41
   Driving Muscle Growth with Peptides   41
   Speeding Up Post-Workout Recovery   49
   The Role of Peptides in Wound Healing   51

3. BRAIN BOOST: ELEVATE YOUR MIND
   WITH PEPTIDES   57
   How Do We Measure Brain Health?   58
   Peptides: A Cognitive Enhancement Tool   67
   Achieving Superior Brain Health and
   Cognition   72
   Integrating Peptides with Other Brain-
   Boosting Strategies   75

4. ENHANCING LONGEVITY AND
   SEXUAL VITALITY WITH PEPTIDES   81
   The Science of Aging   82
   Peptides and Cellular Regeneration   87

Peptides for Longevity and Age-Defying
Benefits   94
Igniting Passion: Peptides and Sexual
Enhancement   96

5. MAKING PEPTIDE THERAPY WORK
   FOR YOU   99
   Why You Should Consider Peptide
   Therapy   101
   What To Expect When Starting Peptide
   Therapy   104
   Risks and Side Effects of Peptide Therapy   106
   Anti-Doping Considerations   108

6. BUILDING YOUR PERSONAL PEPTIDE
   PROTOCOL   111
   Why Individuality Matters in Peptide
   Therapy   112
   Creating a Personalized Peptide Regimen   116
   Advanced Strategies for Optimized
   Peptide Therapy   123
   Identifying Reliable Sources of
   Information   137
   Your Peptide Therapy Checklist   139

7. PEPTIDES MEET BIOHACKING: A
   SYNERGY FOR SUPERIOR HEALTH   145
   What Exactly is Biohacking?   146
   How Biohackers are Using Peptides to
   Get Healthier   151
   The Biohacking Culture   154

8. PEPTIDES: USHERING IN A NEW ERA
   OF WELLNESS   159
   The Latest in Peptide Science   160
   The Exciting Future Uses of Peptides   163
   Proactive Health: Peptides in Preventive
   Care   166
   Pioneers in Peptide Research   168

Glossary of Peptides 173
Conclusion 179
References 183

A NOTE FROM LUMINARY LIFE TO
OUR VALUED READERS

Dear Reader, at Luminary Life, we aim to illuminate the path to better health and well-being. As you delve into the pages of this book, please remember that the insights and suggestions shared are based on our research and understanding of therapeutic peptides. They're designed to inform and inspire, but they shouldn't replace the expertise of a medical professional.

Every individual's health journey is unique. While we've worked diligently to provide accurate and valuable information, this book should not be seen as a definitive guide to diagnose, treat, or prevent any health condition. We strongly encourage you to consult with your physician or a healthcare specialist before making any health decisions or embarking on new treatments.

The world of peptides offers exciting potential, but like any health journey, navigating it with caution and care is crucial. Use the knowledge in this book as a foundational step, and always prioritize expert advice tailored to your specific needs.

We're honored to be a part of your pursuit of better health. Remember, your well-being is paramount, and the best decisions are those made with knowledge and discernment.

To a brighter, healthier future,

Luminary Life

INTRODUCTION

Have you ever felt at odds with your own body and mind, yearning for a wellness breakthrough that always seems just beyond reach? Despite your dedication at the gym and commitment to a strict diet, you may still find that perfect transformation elusive. Slowed mental agility, lingering injuries, or stubborn health issues might be some of your challenges. But what if the answers aren't in another gym session or dietary trend but within your own body?

As a staggering 54.4 million adults in the U.S. face arthritis, it becomes evident that we need innovative treatments that not only alleviate symptoms, but get to the heart of these health challenges. Welcome to the world of therapeutic peptides.

Conventional medicine often focuses on symptom relief, sometimes missing the underlying causes. It's like trying to save a sinking boat without fixing the hole. This might resonate with you, especially if traditional health solutions haven't brought the improvements you desire. Whether it's a nagging injury, a chronic health condition, or diminishing cognitive abilities, you're searching for more.

Perhaps for you, it is an injury that has refused to heal. You dread gym sessions because of the risk of injury. You know that it could take a long time to heal and that long recovery time sets you back on the progress you have made toward your goals. Maybe you are dealing with a chronic health issue that traditional medicine has not addressed. Perhaps you've observed that your mental sharpness, memory, and focus have either declined or plateaued as you've aged. You're seeking an option to break through this cognitive standstill.

The search for effective, scientifically-backed health solutions can be overwhelming. Whether you're navigating the maze of conflicting health advice, feeling let down by conventional medicine, or grappling with the concerns that come with aging, the quest for improved mental clarity, physical vitality, and an extended healthy lifespan is one many embark upon, but few feel they genuinely solve. And that's exactly why peptides hold such promise. They offer the potential for targeted, practical approaches that speak directly to these concerns.

Therapeutic peptides offer a compelling solution to these health challenges. Unlike traditional approaches that merely treat symptoms, often when it's too late, peptides target the underlying causes of illnesses and diseases. This book provides a comprehensive guide to what many consider the future of regenerative medicine. By reading it, you'll gain insights into how to:

- **Understand the science of peptides:** You will be able to easily navigate the complex world of therapeutic peptides as we break down the science into easy-to-understand and relevant material.
- **Discover advanced ways to improve your health:** You will learn new ways to improve your physical strength, mental abilities, and general health that go beyond traditional health practices.
- **Use peptides to build muscle:** You will learn about the peptides that can help with muscle growth and healing, giving you a natural advantage in your quest for a leaner, more muscular body.

- **Boost cognitive performance:** You will explore how peptides can enhance memory, focus, and overall cognitive performance, offering you tools to excel in your professional, personal, and creative pursuits.
- **Understand the role of peptides in aging and longevity:** You will gain insights into the potential of peptides in combating age-related issues and promoting longevity, empowering you to take control of your aging process.
- **Implement peptide therapy safely and effectively:** You will get actionable advice on safely incorporating peptides into your health regimen, with clear guidelines and precautions.

By delving into the rising trends of peptide therapy and biohacking, this book places you at the forefront of a burgeoning health and wellness field. The impact of therapeutic peptides is profound; they're reshaping the worlds of medicine, health, and healing. Their regenerative properties remedy those adversely impacted by contemporary diets and inactive lifestyles. After years of resorting to temporary solutions like diet pills, steroids, and even plastic surgery, many are embracing a fresh alternative: at-home peptide injections.

Prominent personalities, such as Upgrade Labs CEO Dave Asprey, have openly shared their experiences with peptide use, bringing this approach into the spotlight. Similarly, renowned fitness expert and biohacker Ben Greenfield lauds the efficacy of therapeutic peptides. He attributes the swift recovery of a stubborn Achilles tendon injury to peptides, enabling him to return to his rigorous training regimen swiftly. The allure of peptides is not limited to the fitness sphere. Hollywood stalwart Sylvester Stallone, celebrated for his portrayal of Rocky Balboa, champions peptides, crediting his exceptional physique and longevity to a meticulous peptide routine.

However, it's essential also to discuss potential risks and ethical considerations. As with any therapeutic intervention, peptides can have side effects, and misuse may lead to unwanted repercussions. Especially for athletes, it's imperative to understand the position of sports regulatory bodies on peptides. Several peptides are listed as prohibited substances by anti-doping agencies, potentially resulting in penalties for users. Throughout this book, we'll delve into these concerns, ensuring you are well-equipped to make prudent decisions.

But when used responsibly and with understanding, the possibilities are immense. Imagine the empowerment of optimizing your health or finally achieving that lean, muscular physique you've been striving for. This book serves as a guide to understanding how peptides can help you meet various goals, from boosting immunity and losing fat to building muscle, enhancing cognition, slowing aging, and healing injuries. Instead of relying on generic health tips or traditional fitness strategies, this book offers cutting-edge, science-based solutions for those seeking remarkable outcomes. The content balances scientific data and practical, easily understandable advice, making it an invaluable resource for a broad spectrum of readers.

Toward the end of the book, you'll also find a glossary encompassing the peptides and their analogs that we've discussed. While it doesn't capture every peptide out there, we aim for it to be a handy reference for you. So, whenever you need a quick refresher on any peptide topic we've delved into, you know where to turn.

After reading this book, you will be equipped to make informed decisions about your health and well-being, potentially leading to enhanced physical vigor, cognitive capabilities, and faster recovery times. This knowledge will empower you to transcend your health and performance levels, bringing you closer to your ultimate goal of peak human potential.

About Luminary Life

Luminary Life is an avant-garde group committed to advancing and promoting optimized mental and physical performance. Our approach blends biology, neurology, and fitness, allowing us to explore the captivating intersection of science and human potential.

We have forged a deep understanding of the human body and mind, allowing us to explore methods that enhance physical vigor, cognitive capabilities, and overall well-being. We believe that every person holds the key to unlocking their fullest potential. Our mission revolves around equipping you with the tools and knowledge necessary to seize control of your biology.

Drawing from our expertise and personal passion for biohacking, our commitment is to guide you toward self-improvement and optimal health. As we delve into the intricacies of peptides, we hope to equip you with not just knowledge but tools for tangible transformation. With the foundation now laid, let's embark on our first deep dive into peptides and discover the potential that awaits.

1

UNDERSTANDING PEPTIDES AND THEIR ROLE IN THE BODY

Imagine your body as a grand orchestra, where every cell plays its unique part in the symphony of life. In this analogy, peptides serve as the expert conductors, directing each cell to work harmoniously and produce the best possible performance for your overall health. This chapter will delve into how peptides guide the intricate workings of your body, from the tiniest cellular components to the mechanisms that regulate your mental processes. In essence, peptides are the key players in optimizing your body's functions.

By the end of this chapter, you will understand what peptides are and how they function within the tapestry of your body's systems. You will know their remarkable potential as an agent of healing, vitality, and longevity.

As you delve into the study of peptides, you'll gain insights into the intricate cellular interactions that underlie your overall health and well-being. This scientific exploration reveals how peptides are crucial in regulating these cellular processes, forming the foundation of your body's optimal functioning.

BASICS OF PEPTIDES

Peptides are small yet essential components that contribute to the proper functioning of our bodies. Composed of smaller units called amino acids, peptides serve as building blocks for various biological processes. Amino acids are the fundamental elements that our body uses to form proteins. A peptide consists of a sequence of 2 to 50 amino acids linked together by peptide bonds. The name given to a peptide depends on the number of amino acids it contains: a chain with 10 to 20 amino acids is termed an oligopeptide. A chain with more than 20 amino acids is called a polypeptide.

Each amino acid within the peptide is referred to as a 'residue,' a term that signifies what remains after a dehydration reaction has occurred to link the amino acids together. Peptide formation involves three main steps. First, amino acids are prepared for synthesis. Next, they are sequentially linked to form a longer chain. Finally, additional chemical processes are carried out to ensure the peptide chain is complete and functional. These peptides serve as the foundational elements for creating proteins.

The peptide bonds that link amino acids in a chain are highly stable and resistant to breakdown, even under extreme conditions such as high temperatures or the presence of concentrated chemicals. All peptide amino acids have a similar molecular structure, featuring positive and negative regions. Due to their role as foundational elements of proteins, peptides can be synthesized in a laboratory setting. These synthetic peptides can regulate specific cellular activities and participate in numerous biochemical processes within the body, offering a wide range of potential applications for addressing health concerns.

Peptides and proteins serve essential functions in biology and are composed of amino acid chains. However, they differ primarily in size and structure–proteins are larger and usually consist of more than 50 amino acids. Proteins often have more complex, well-defined structures and are fundamental to cell function, providing cellular shape and responding to external signals. On the other hand, peptides usually have simpler, less defined structures and primarily serve to regulate the activities of other molecules.

Proteins and peptides differ not only in size but also in function. Proteins are complex molecules formed from one or more polypeptide chains, whereas peptides are smaller and contain fewer amino acids. Due to their smaller size, peptides are generally more easily absorbed by the body than proteins. They can more readily penetrate tissues such as the intestines and skin, allowing quicker entry into the bloodstream. Peptides used in supplements are commonly derived from various protein sources, including animal-based proteins like meat, fish, and milk and plant-based proteins like soy and quinoa.

Bioactive peptides are specialized peptides that have distinct health benefits. These peptides are not all the same; they can differ in structure and function, which means they have varying effects on the body. For example, collagen peptides are primarily used to improve skin health and slow aging. On the other hand, creatine peptides are commonly used to help build muscle mass. Some peptides are even formulated to boost athletic performance. Bioactive peptides serve different health-related purposes depending on their unique properties and formulations.

Building on the diverse roles of bioactive peptides, scientific research has highlighted their potential in anti-aging treatments, mainly through their influence on collagen. Collagen is a vital protein found in various body parts, including the skin, nails, and hair. Collagen peptides, which are more easily absorbed by the body than whole collagen proteins, are created by breaking down these larger molecules. Multiple studies suggest that supplements containing collagen peptides can effectively reduce skin wrinkles and improve skin hydration and elasticity. Additionally, some peptides are known to stimulate melanin production, a pigment that helps protect the skin from sun damage. As a result, peptides are commonly found in anti-aging skincare products.

Beyond the skin, collagen plays a pivotal role in overall health, especially in wound healing and bone density. Collagen peptides have been shown to reduce inflammation and exhibit antioxidant properties, thereby enhancing the body's natural healing process. Furthermore, peptides may play a role in mitigating age-related bone loss. Animal studies have demonstrated that a moderate intake of collagen peptides, combined with exercise, increased bone mass in rats. This suggests that peptides could be a potential treatment option for addressing bone loss due to aging.

While more research is needed to elucidate the impact of peptides on bone loss fully, existing studies do indicate a positive effect on muscle strength and mass. One study involving older adults showed that the combination of collagen peptide supplements and resistance training resulted in significant improvements in muscle strength and mass within a four-week period.

Collagen peptides are not the only peptides with potential muscle-building benefits; creatine peptides are also commonly used for this purpose. These peptides are thought to be more easily digested by the body, and anecdotal reports suggest they may lead to fewer digestive issues. Generally, healthy individuals are unlikely to experience side effects from peptide supplements, as these substances are structurally similar to peptides found in a typical diet. However, it's important to note that some oral peptide supplements may not be fully absorbed into the bloodstream due to limitations in the body's ability to break them down into individual amino acids.

In a specific study examining the potential side effects of peptide supplements, 30 female participants were administered oral collagen peptide supplements for a duration of two months, with no adverse reactions reported. Some individuals have reported skin irritation, such as itching and rashes, from using topical peptide creams. Therefore, purchasing products from reputable sources is advisable, and discontinue use if adverse reactions are observed.

DIVE INTO PEPTIDE BIOCHEMISTRY

Amino acids are the fundamental building blocks that combine to form peptides and proteins. When these amino acids link together, they form peptide bonds. These bonds are created inside the cell by tiny factories called ribosomes. The specific order in which amino acids are connected is directed by a molecule called mRNA, which acts like a recipe for the ribosome to follow. Due to the nature of peptide bonds, peptides have unique characteristics that influence their structure and how they fold. Once formed, peptides play diverse roles in the body, such as aiding digestion or facilitating communication between cells.

As the ribosome connects amino acids together, a reaction takes place, releasing water and solidifying their bond. This growing chain is termed a polypeptide and is a primary component of proteins. After its initial formation, the polypeptide may undergo further adjustments to reach its final and functional form. This can involve trimming certain sections, and once refined, the peptide gets packed into transporters known as vesicles. These vesicles carry the peptide to another part of the cell called the Golgi apparatus for any final modifications. Once fully prepared, the peptide embarks on its specific mission within or outside the cell.

In the lab, scientists can also create what are known as synthetic peptides. They use solid-phase peptide synthesis to assemble amino acids in a specific order, similar to how our cells do it. One common type of these lab-made peptides is peptide hormones.

Peptide hormones are chains of amino acids that function as signaling molecules in the body. Depending on their composition, they can range from just a few amino acids in length to over a hundred. Produced in specific glands or tissues, they travel through the bloodstream to exert their effects on target organs or cells. Their roles encompass various biological processes, from growth regulation to metabolic control. Examples of peptide hormones include:

- **ADH (Antidiuretic Hormone) and oxytocin** are peptide hormones produced in the brain, specifically in the hypothalamus. While their structures are similar, they differ slightly in their amino acid composition. ADH plays a crucial role in regulating water balance in the body, while Oxytocin has various functions, including roles in social bonding and childbirth.
- **Insulin** is a peptide hormone composed of 51 amino acids. It plays a crucial role in regulating blood sugar levels by facilitating the entry of

sugar into cells. Essentially, insulin acts like a key, unlocking cells to absorb sugar from the bloodstream, helping maintain balanced blood sugar levels.
- **Glucagon** is a hormone produced from the precursor molecule proglucagon. It is released by specialized cells in the pancreas, primarily when blood sugar levels are low or when there is an excess of amino acids in the body. Glucagon plays a vital role in maintaining metabolic balance by promoting the conversion of stored energy into usable sugar, thereby elevating blood sugar levels.
- **The pro-opiomelanocortin (POMC)** gene family produces a large precursor protein that is then cleaved into more minor, active hormones and peptides. These include melanocyte-stimulating hormone (MSH), adrenocorticotropic hormone (ACTH), beta-lipotropin, and beta-endorphin. These components are synthesized in both the brain and peripheral tissues and serve various bodily functions.
- **Secretin** is a peptide hormone consisting of 27 amino acids, originating from the SCT gene as a precursor called prosecretin. When exposed to gastric acid in the stomach, prosecretin is

converted into its active form, secretin. The active secretin stimulates the pancreas and bile ducts to release bicarbonate, which neutralizes stomach acid. In this way, secretin acts as a protective mechanism against the damaging effects of excess stomach acid.

- **Calcitonin gene-related peptide (CGRP)** is a peptide hormone composed of a 37-amino acid chain. It is present in specific nerve tissues and affects the central and peripheral nervous systems. Research suggests that CGRP may influence energy levels and appetite. Additionally, there is evidence to indicate a potential link between CGRP and the occurrence of migraines.
- **Natriuretic peptides** are produced by heart cells in response to cardiac stress. The family includes atrial natriuretic peptide (ANP), brain natriuretic peptide (BNP), and C-type natriuretic peptide (CNP). These peptides serve protective roles for both the heart and kidneys by inhibiting excessive heart growth and dilating blood vessels. ANP consists of a 28-amino acid chain, BNP comprises 32 amino acids, and CNP contains 22 amino acids.

Peptides are small but powerful molecules that play crucial roles in various bodily functions, from helping wounds heal to fighting off infections. These molecules are produced by our bodies and interact with specific proteins or receptors, like keys fitting into locks. When they attach, they can trigger a chain reaction of cellular events. If something goes wrong in this process, it can result in problems such as excessive cell growth or even tumors.

Researchers are investigating the potential of peptides in cancer treatment and diagnosis. In the lab, scientists can create synthetic peptides that act like markers to pinpoint tumors during imaging tests. This technique could revolutionize how we detect and treat cancer.

Peptides are also our body's first line of defense against infections. Known as antimicrobial peptides (AMPs), these molecules are produced in areas like the skin and specific blood cells. They are designed to neutralize harmful bacteria and can operate in various conditions, from salty environments to different acidity levels. However, some bacteria have developed mechanisms to resist these natural defenses.

In wound healing, peptides direct cells to perform specific tasks, such as removing bacteria and promoting tissue repair. For example, a peptide known as syndecan activates substances that help repair damaged tissues.

Peptides also support our skin's natural defenses against bacteria and viruses. However, things can go awry in conditions like atopic dermatitis, where the skin's defense mechanisms are weakened, or in rosacea, where an overproduction of a specific peptide causes inflammation.

Overall, peptides have diverse and significant roles in our body, and ongoing research aims to harness their potential for medical applications.

CLASSES OF PEPTIDES

Based on their properties and functions, peptides are divided into groups. Some of the most common groups include:

- **Neuropeptides**

Neuropeptides are tiny molecules made by nerve cells to help regulate communication between cells. They are processed into their active forms, which can produce different versions of neuropeptides. They mainly belong to two groups: the secretin family and the rhodopsin-like family. While serotonin and dopamine are often associated with neural signaling, it's important to note that they are not neuropeptides but rather neurotransmitters.

- **Antimicrobial peptides**

Known as host defense peptides, these molecules aid the immune system in responding to threats. They can be created through two different processes: non-ribosomal or ribosomal synthesis. Fungi and bacteria mainly produce non-ribosomal peptides. These peptides are complex and often include multiple components. Well-known examples are penicillin and vancomycin. On the other hand, ribosomally synthesized antimicrobial peptides are made by a wide range of organisms, including plants, animals, and even some viruses and bacteria. Examples of these include polymyxin B and bacitracin.

- **Bacterial peptides**

Bacterial peptides are unique molecules made by bacteria. These molecules can serve various functions and come in different types, such as lipoproteins, flagellar peptides, and enterotoxins. These peptides can be made by different kinds of bacteria, some of which are more resistant to antibiotics than others. The peptides can be either positively charged or have no charge at all. Examples of these bacterial peptides include mersacidin and nisin.

- **Anticancer peptides**

Anticancer peptides are specialized molecules that target and kill cancer cells. They mainly contain amino acids like leucine, lysine, and glycine. These peptides are a significant focus in cancer treatments because they are highly selective and can be easily modified for different uses. They work by creating holes in the cell membranes of cancer cells, leading to their destruction.

There are three main types of anticancer peptides based on how they work. The first type directly attacks cancer cells, causing them to die or stop growing. The second type attaches to drugs and helps deliver them to the cancer cells. The third type works indirectly by activating other systems in the body, like hormone receptors or immune responses, to target and kill the cancer cells.

- **Cardiovascular peptides**

Cardiovascular peptides are essential for normal and abnormal heart conditions and blood vessels. They regulate blood pressure, manage heart diseases such as coronary artery disease and congestive heart failure, and control hypertension. Some of these peptides work by lowering blood pressure and expanding blood vessels (vasodilation), while others help to reduce the abnormal growth of heart tissue (ventricular hypertrophy). A few even have a role in modifying the activity of specific genes, which can also affect blood vessel expansion.

- **Antifungal peptides**

Antifungal peptides are molecules specifically designed to fight against fungal infections. These peptides target fungi that can cause diseases in both humans and plants. Researchers isolate these peptides to study how they combat fungal growth and infection.

- **Opiate peptides**

Opiate peptides are involved in regulating systems in the body related to stress and pain relief. They achieve this by interacting with specific receptors, leading to various physiological and pharmacological outcomes. Some opiate peptides even have properties that counteract the effects of opioids, essentially acting as antiopioids.

- **Endocrine peptides**

Endocrine peptides are produced within the endocrine system and are stored in membrane-bound compartments called secretory vesicles. Due to their water-soluble nature, they cannot easily cross cell membranes. Instead, they interact with specific receptors on the surface of target cells to exert their effects. For example, adiponectin is an endocrine peptide that promotes fatty acid breakdown and facilitates glucose absorption in skeletal muscles.

- **Plant peptides**

Plant-derived peptides are molecules sourced from plants that have demonstrated health benefits for humans. These peptides have been shown to reduce cholesterol levels and lower blood pressure. Additionally, they possess anti-inflammatory and anti-cancer properties and can regulate the immune system.

- **Venom peptides**

Venom peptides are toxic substances produced by certain animals, such as scorpions, snakes, spiders, and snails. These peptides serve as a defense mechanism against predators or assist in capturing prey. They bind to specific receptors in the target organism, triggering various physiological responses.

THERAPEUTIC PEPTIDES: THE NEXT WAVE IN HEALTH AND FITNESS

In the field of drug development and pharmacology, compounds are usually categorized as either small molecules or large molecules, also known as biologics, based on their structure and how they function in the body. Small molecules are simpler and can be easily synthesized in a laboratory. They are often administered orally and act quickly, but their effects may be short-lived. Common painkillers serve as examples. On the other hand, biologics are more complex molecules, frequently proteins, derived from living cells. They usually require injection and are effective in treating specific conditions because they closely resemble the body's natural molecules.

Peptides are gaining attention for their unique position between these two categories. Like small molecules, they can be produced in a lab and act quickly. However, like biologics, they are composed of amino acids and tend to be well-tolerated by patients. The U.S. Food and Drug Administration (FDA) defines a molecule with more than 50 amino acids as a biologic and those with fewer as peptides. Peptides often produce long-lasting effects, similar to biologics.

This hybrid nature of peptides makes them versatile for treating various diseases and conditions. Consequently, the pharmaceutical industry increasingly focuses on peptide-based drugs, which often have higher success rates in clinical trials than small molecules. One of the first and most well-known therapeutic peptides is insulin, which has been used in diabetes treatment for over a century. As of 2022, according to the FDA, around 100 therapeutic peptide products were available in the U.S., Japan, and Europe, generating global sales of $20 billion. This market is expected to grow at a rate of 8.2% leading up to 2026.

Outside of clinical applications, peptides are also becoming popular in the health and wellness community. For example, Ben Greenfield, a competitive athlete and fitness coach, uses peptides for various health goals, such as muscle gain and accelerated healing. Margaret Josephs, from the reality television series The Real Housewives of New Jersey, credits peptide therapy for her weight loss and hormone balance. Actress Jennifer Aniston also reports benefits from peptide treatments for anti-aging.

In the bodybuilding community, peptides stimulate the production of growth hormones, helping people build muscle and lose fat more effectively. Overall, the versatility and efficacy of peptides in diverse applications make them a growing focus in pharmaceutical research and general health and wellness.

HARNESSING THE POWER OF PEPTIDES

In chemistry, peptides serve critical functions in various research applications, including protein modification and purification. These scientific uses extend into practical applications, particularly in creating medications and health supplements. Researchers from diverse fields—from medicine to food science to cosmetics—are increasingly investigating the potential of peptides due to their potency, specificity, and safety profile. These characteristics position peptides as promising agents that fill the gap between small molecular drugs and more complex biologics.

A key advantage of peptides is their natural affinity for specific receptors in the human body. They can engage these receptors to initiate targeted physiological responses, making them valuable drug discovery and delivery tools. For example, in cancer diagnostics and treatment, peptides can be engineered to bind to specific receptors overexpressed in tumor cells, aiding in accurate imaging and potentially targeted therapies.

Peptides are also becoming increasingly recognized in functional foods and nutraceuticals, which are products derived from food sources that provide extra health benefits in addition to their essential nutritional value. Derived from various food sources, these peptides exhibit therapeutic properties, offering health benefits that complement a well-balanced lifestyle.

However, it's important to note that while peptides show promise, their misuse or unregulated usage could lead to adverse effects. For instance, peptides designed for muscle growth are relatively new to scientific research, and comprehensive data on their long-term safety is still limited.

Peptides are distinct from secretagogues, although some peptides can function as such. Specifically, secretagogues are substances that trigger another substance to be secreted. A pertinent example in this context is growth hormone secretagogues, which stimulate the release of growth hormone. While short-term usage of these compounds is generally well-tolerated, some studies have raised concerns. A review indicated that growth hormone secretagogues could elevate blood sugar levels and decrease insulin sensitivity. One specific secretagogue, MK-677, known as a "ghrelin mimetic," has potential side effects like fluid retention, muscle pain, and increased appetite. Due to these potential risks, it's advisable to use peptide-based therapies under the guidance of healthcare professionals.

Overall, peptides are emerging as potent, versatile molecules in both scientific research and practical applications, although their long-term safety profile still requires further investigation. In the next chapter, you will learn how they help improve muscle mass and speed up post-injury recovery. You will understand how you can use them to improve your athletic performance.

DISCLAIMER

Before delving into Chapter 2, it is crucial for readers, especially those involved in competitive sports, to be aware that substances such as human growth hormone (HGH) and insulin-like growth factor (IGF-1) are generally prohibited by many anti-doping agencies worldwide. If you are an athlete or are involved in any form of competitive sport, it is your responsibility to thoroughly investigate and understand the specific regulations and restrictions related to these substances within your sport and jurisdiction. Failure to do so can lead to severe penalties, including disqualification and bans. This book provides information for educational purposes, and it is the reader's obligation to ensure compliance with all applicable rules and regulations.

2

MUSCLES, RECOVERY, AND THE MAGIC OF PEPTIDES

Consider the possibility of your body recovering rapidly from workouts. Envision building lean muscle efficiently and healing from injuries with enhanced speed. While this might sound ambitious, therapeutic peptides offer such potential benefits. In this chapter, we will delve into how peptides can expedite recovery and enhance athletic performance. Furthermore, guidance from fitness professionals will offer practical applications for peptide therapies in your regimen.

DRIVING MUSCLE GROWTH WITH PEPTIDES

Many bodybuilders, athletes, and fitness enthusiasts are turning to peptides to enhance muscle growth, reduce fat, and boost appetite during bulking phases. In this context, growth hormone secretagogues are of primary interest. Two main categories of these secretagogues are relevant to peptides: growth hormone-releasing peptides (GHRPs) and growth hormone-releasing hormone (GHRH) agonists. GHRH agonists are peptides that mimic the effects of GHRH, leading to increased secretion of growth hormone. When referencing peptides in bodybuilding, these specific types —GHRPs and GHRH agonists—are typically the focal point due to their role in promoting growth hormone release. Although there's a wide variety of peptides with different benefits, it's vital to note that not all operate through growth hormone-related pathways. The emphasis of this section is on those that do.

Specific peptides, such as CJC-1295, are commonly utilized by bodybuilders aiming to enhance fat loss and muscle growth. They work by stimulating the production of HGH (human growth hormone) and its derivative, insulin-like growth factor 1 (IGF-1). Though HGH directly leads to the creation of IGF-1, the two have distinct physiological effects on the body. Notably, the brain plays a pivotal role in regulating HGH production: the hypothalamus releases GHRH, which then triggers the secretion of growth hormone into the bloodstream.

HGH plays a crucial role in mobilizing fatty acids, enabling the body to utilize them for energy. This aspect of HGH benefits those aiming for fat loss while preserving lean muscle mass. However, while HGH does have some direct anabolic effects, it significantly amplifies its growth-promoting actions by stimulating the liver to produce IGF-1, a potent anabolic hormone. Thus, IGF-1 often emerges as the more influential factor for muscle growth, given its central role in promoting the uptake of amino acids and their incorporation into muscle proteins.

The body primarily produces growth hormone during sleep in a pulsatile manner. As individuals age, especially entering their 30s, growth hormone production naturally decreases. HGH peptide therapies are designed to augment this hormone's production when it starts to diminish. Although some athletes and bodybuilders opt to use both IGF-1 and HGH, these peptides can be costly. As a result, many turn to more cost-effective options like GHRPs and GHRH analogs.

Consider the body as an intricate system akin to a construction site. In this analogy, muscles represent the structures, proteins serve as the building materials, and peptides function as the builders. When individuals exercise, they induce minor muscle damage (known as micro-tears). Peptides then play a pivotal role in repairing these damages, analogous to builders mending structures. Therapeutic peptides amplify this repair process, bolstering muscle strength and recovery. Their primary function is to boost the production of specific hormones vital for muscle development and optimal athletic performance, leading to muscle enlargement or hypertrophy. Unlike some performance-enhancing drugs such as steroids, peptide supplements are rapidly assimilated by the body and generally present fewer side effects.

Peptides that stimulate the growth hormone's production have interested researchers for decades. In the 1970s, scientists began exploring synthetic analogs of enkephalin, a naturally occurring peptide in the brain, to better understand the role of the growth hormone. They selected enkephalins due to their small size and structural similarity to opiates, which are known to trigger the release of the growth hormone. At that time, scientists had not fully understood the nature of growth hormone-releasing hormone (GHRH). However, they suspected a link between natural peptides, such as enkephalins, and GHRH. While it was widely believed that opiates stimulated growth hormone release primarily through the hypothalamus, scientists theorized that certain natural peptides could trigger the hormone through both pituitary and hypothalamic pathways. To investigate this possibility, they conducted lab tests on specific variants of these peptides, known as leucine-enkephalins and methionine-enkephalins, as well as their related compounds.

Further studies revealed that a particular pentapeptide, DTrp2, when attached to methionine-enkephalin, could spur the growth hormone's release in vitro, albeit with certain constraints. This specific pentapeptide drew considerable attention in ensuing research. Its defined amino acid sequence distinctly nudged the pituitary gland to discharge the growth hormone while simultaneously not activating opioid receptors. Such precision meant that other hormones produced by the pituitary gland, like luteinizing hormone (LH) and prolactin (PRL), weren't simultaneously released.

However, when the pentapeptide was tested on live subjects (in vivo), it did not enhance the growth hormone production as anticipated. Originally thought to influence the pituitary gland, this peptide was classified as a peptidomimetic, meaning it was believed to mimic the function of a natural peptide. Researchers hypothesized that by modifying its amino acid structure, they might be able to design a more effective version of the peptide.

In their continued quest to identify effective peptides, researchers eventually developed new analogs that are effective both in vitro and in vivo. A notable outcome of this research was the development of a hexapeptide (six amino acids linked together) known as GHRP-6. Initially, it was thought that these GHRPs worked by mimicking the action of a natural hormone in the hypothalamus responsible for triggering the release of the growth hormone. However, the synthetic nature of GHRPs, marked by uncoded amino acid residues, set them apart from natural peptides. As research progressed, these GHRPs underwent further refinement, leading to the understanding that they function as peptidomimetics, meaning they imitate the functions of natural peptides but have distinct structural differences.

In 1982, scientists successfully isolated natural GHRH from a tumor that led to acromegaly, a condition characterized by an overproduction of the growth hormone, in a patient. The GHRH agonist sermorelin was developed after this discovery. Originally intended as a weight loss drug, it also addressed growth hormone deficiencies. Today, a variety of GHRHs and GHRPs are available. However, the extent to which the body's natural GHRH, whose levels can be elevated by GHRPs, contributes to the release of the growth hormone remains an open question.

Studies involving animals and humans have shown that GHRPs don't increase the body's natural levels of

GHRH when given in low doses. In these situations, the body's own GHRH seems to play a secondary role in the growth hormone release initiated by GHRPs rather than being the main trigger. However, when GHRPs are given in high doses, they do seem to raise the body's natural GHRH levels, suggesting that in these cases, GHRH plays a more direct role in the growth hormone release caused by GHRPs. Additionally, recent research on young men found that blockers of GHRH can actually stop the growth hormone response that GHRPs usually induce.

A widely used peptide combination for weight loss and muscle development is the GHRH agonist CJC-1295 and ipamorelin, a GHRP. Typically, they are administered in a 1:1 ratio based on micrograms per kilogram of body weight. For instance, an 80kg individual would take 80 mcg of CJC-1295 and 80 mcg of ipamorelin per dosage. This combination is known to elevate growth hormone levels in the body quickly. It's crucial to administer these peptides via intramuscular injection to maintain their efficacy, as oral consumption exposes them to digestive enzymes, disrupting their amino acid sequence and nullifying their benefits. Another option for administration is through nasal delivery.

In most situations, achieving significant benefits requires consistent use and multiple doses of peptides.

Short-term usage might mainly result in an increased appetite without other noticeable effects. For weight loss, it's recommended to administer the peptides approximately 30 minutes before cardiovascular exercises. This timing leverages the rise in growth hormone levels, promoting fat breakdown and the burning of fatty acids. For muscle gain, injecting the peptides roughly 20 minutes before eating is advisable.

As previously detailed, growth hormone secretagogues are among the most potent peptides for muscle development. However, creatine peptides and collagen peptides also show promise based on their applications. Creatine is an amino acid stored in the muscles, and it offers numerous potential advantages when paired with proper exercise. Supplements containing creatine are widely used to improve energy utilization. These supplements can enhance strength, elevate energy levels, and promote muscle growth. Their efficacy as exercise supplements can be attributed to their rapid absorption and gut-friendliness.

Collagen peptides are another option for muscle growth, though they may not be as immediately effective as growth hormone secretagogues. Beyond muscle enhancement, collagen peptides can support healthier, more hydrated skin. Combining them with resistance training is essential to optimize their effectiveness for muscle development. A study involving young men showed that those who took a collagen peptide supplement experienced more significant gains in body mass and muscle strength than those who only engaged in resistance training.

SPEEDING UP POST-WORKOUT RECOVERY

Gone are the days when only elite athletes had concerns about optimal recovery and peak performance. Today, even a casual runner has methods to manage fatigue and soreness. A comprehensive fitness routine should include a recovery strategy. Injuries become more likely without proper muscle rest, healing, and regeneration. In addition to injury prevention, effective recovery contributes to muscle growth and strength development. Although recovery traditionally relies on nutrition, sleep, and time, other options are available.

Peptides can significantly enhance recovery. They amplify the body's natural tissue repair mechanisms and expedite the recovery process. Properly selected peptides can boost blood circulation to muscles, fostering rapid healing. Furthermore, peptides can guide cellular structures and essential proteins to areas needing repair, accelerating the recovery process. Studies indicate that peptides can promote better sleep and reduce post-workout discomfort, further aiding recovery. As a result, individuals can return to their training or physical activities more swiftly.

Many peptides can help you in the recovery process, but these four are the most common:

- **Mechano growth factor (MGF):** MGF is recognized for its role in muscle development. It stimulates damaged cells to divide and form new ones, contributing to muscle repair and growth.
- **CJC-1295:** This peptide facilitates cell regeneration and repair. An added benefit of its mechanism is a reduction in body fat.
- **Thymosin beta-4:** This peptide accelerates tissue repair by enhancing blood circulation to injured tissues, leading to quicker recovery.
- **IGF-1:** This naturally occurring growth factor in the body aids tissue regeneration through a mechanism involving the pituitary gland.

If you are dedicated to training or frequently experience soreness post-exercise, incorporating peptides into your recovery regimen can offer enhanced benefits compared to traditional methods alone. Prioritizing peptide-assisted recovery underscores a commitment to optimizing both training and recovery.

Many individuals often conflate peptides with steroids, likely due to their shared association with enhancing muscle growth. However, their mechanisms of action are distinct. Anabolic steroids function by mimicking the effects of testosterone, a naturally occurring sex hormone responsible for muscle development. On the other hand, peptides act on the pituitary gland to encourage the body's natural hormone production.

Injecting steroids directly increases hormone levels in the bloodstream. In contrast, peptides stimulate the body's natural hormone production process, leading to a rise in certain hormones like growth hormone and testosterone. Steroid use is associated with harmful side effects, including increased aggression, cardiovascular damage, heightened risk of testicular cancer, and reduced sperm count. Recent studies indicate that peptides do not exhibit these detrimental effects, positioning them as a safer alternative to steroids.

THE ROLE OF PEPTIDES IN WOUND HEALING

In clinical settings, chronic wound infections paired with antibiotic resistance pose significant challenges, impacting millions worldwide. Research into peptides for addressing these issues revealed benefits extending beyond wound care. Peptides can accelerate specific wound healing mechanisms by enhancing cell growth and movement, regulating immune responses, promoting blood vessel formation, converting fibroblasts into myofibroblasts, and facilitating collagen accumulation.

In a study investigating the antimicrobial properties of peptides, researchers examined cathelicidin-DM for its potential wound-healing capabilities. They aimed to understand how this peptide influences cell movement and growth. Specifically, they looked at its impact on the MAPK (mitogen-activated protein kinase) signaling pathways, a kind of 'cellular communication system' that helps cells respond to their environment and decide actions like when to grow or divide. Using mice as subjects, the study evaluated the wound-healing effects of the peptide using immunohistochemical assays, which are specialized tests that visualize and pinpoint specific proteins or markers within tissue samples.

Results showed that cathelicidin-DM heals both non-infected and infected wounds via different pathways, presenting a novel approach to treating infected chronic wounds. The peptide was found to boost cell growth in a concentration-dependent manner, activate the MAPK signaling pathway without influencing the secretion of other hormones, and enhance wound healing and collagen deposition in animals.

Naturally occurring peptides play a role in inflammatory responses, specifically in triggering anti-inflammatory actions. They function by suppressing or reducing the activity of inflammatory mediators. Inflammation is the body's defense mechanism against infections and helps repair damaged cells. However, excessive inflammation can lead to various diseases, including skin conditions, asthma, and arthritis. Research has shown that excessive production of inflammatory molecules, such as cytokines and oxidants, can be problematic. Historically, doctors used small-molecule drugs to manage inflammatory disorders, but these have been associated with adverse side effects.

Researchers are currently exploring alternative anti-inflammatory treatments, focusing on peptides. The proper functioning and balance of intricate protein networks are essential at the cellular level. Disruptions in these protein systems can lead to diseases. Moreover, protein-to-protein interactions (PPIs) are fundamental to cell signaling. When these interactions are disrupted, complications arise. Peptides can influence these protein complexes, offering potential treatment options for various disorders.

Numerous naturally occurring peptides possess a wide range of biological properties. While some have either pro-inflammatory or anti-inflammatory effects, others, known as antimicrobial peptides, specialize in targeting and eliminating pathogens. These antimicrobial peptides can combat two main types of bacteria, gram-positive (with a thicker protective layer) and gram-negative (with a thinner protective layer), as well as cancer cells and fungi. They typically function by interacting with cellular surfaces and integrating into the membrane bilayers, forming pores. In some instances, these peptides penetrate the cell entirely and bind to intracellular components, disrupting their metabolic functions and leading to cell death. Properly combined peptides can enhance their binding efficiency, which is especially beneficial for wound healing.

In a study examining Inflammatory Bowel Disease (IBD)–a long-term gastrointestinal condition encompassing both Crohn's disease and ulcerative colitis (UC)–researchers sought alternative treatments due to the limited efficacy of traditional methods. They gave patients peptides, which reduced the activity of macrophages (cells that ingest harmful organisms and secrete both pro-inflammatory and antimicrobial mediators) and decreased inflammation. Specifically, they utilized the peptide chromofungin (CHR: CHGA47-66) for its ability to initiate the release of other bioactive peptides, such as serpinin, which further contributed to the decrease in inflammation.

Researchers have discovered numerous peptides with wound-healing properties and have developed dressings derived from these peptides. Such advancements offer exciting possibilities for treating infected wounds. Often, non-healing wounds are populated with bacteria such as *Enterococcus faecalis, Staphylococcus aureus,* and *Pseudomonas aeruginosa.* In a healthy person, the immune system releases chemicals that regulate these bacteria on the skin and initiate a series of healing processes after an injury.

However, in individuals with conditions like diabetes or venous insufficiency, their immune response may be compromised. This can result in bacterial growth and the formation of biofilms. Biofilms are protective communities of microorganisms, often enveloped in a slimy matrix, that adhere to surfaces and can be difficult to remove. These biofilms are often resistant to antibiotics, making them particularly problematic. Peptides present a potential solution: they can counteract infections due to their antimicrobial properties, support cell migration, and reduce inflammation, aiding the healing process.

Peptides offer a promising avenue in the realm of fitness and recovery. They have the potential to amplify muscle growth and expedite injury recovery. Essentially, they act as small catalysts, propelling muscle development and aiding in achieving a leaner, more muscular physique. This ensures you return to your activities swiftly. This section has equipped you with the underlying science and practical guidance on effectively utilizing peptides for muscle augmentation and injury recovery. You now possess the knowledge to harness these natural enhancers to their fullest. The question remains: will you implement these strategies? Will you elevate your fitness journey? The choice is yours.

Having delved into the benefits of peptides for physical enhancement, let's transition to another intriguing dimension: cognitive optimization.

3

BRAIN BOOST: ELEVATE YOUR MIND WITH PEPTIDES

The ability to effortlessly recall information and sustain concentration is admired by many. But what if you learned that these abilities aren't just innate but can be enhanced through natural methods?

This chapter delves deep into the world of peptides and their profound influence on cognitive functions. Beyond muscle repair and recovery, peptides are pivotal in bolstering brain performance—from sharpening memory to enhancing focus. Through a review of current scientific research, we'll demystify how these small chains of amino acids can reshape our cognitive prowess. By the end of this exploration, you'll be equipped with knowledge of how peptides might offer you a pathway to a sharper mind and improved concentration.

However, it's essential first to understand the complexities of the brain and how its health is gauged. After all, by grasping the significance of brain health, we can better appreciate how peptides might play a transformative role in enhancing cognitive functions and overall neurological well-being. Therefore, let's first explore the intricate workings of the brain and the contemporary methods by which its health is measured.

HOW DO WE MEASURE BRAIN HEALTH?

The brain acts as the central hub for our nervous system, overseeing emotions, movement, memory, and cognition. Its intricate functions are a testament to the intricacies of biological evolution. Prioritizing brain health is crucial for overall well-being and longevity, especially given the rise in neurological disorders with increasing age in the population. Currently, there is no universally accepted definition of brain health. Most definitions provide a broad overview of typical brain functions or emphasize specific aspects of brain well-being.

The US Centers for Disease Control and Prevention defines brain health as the capability to perform all cognitive functions, such as memory, speech, judgment, and learning. On the other hand, the American Heart

Association's presidential advisory characterizes it as the typical performance levels seen in individuals of the same age group without any brain or other organ system diseases affecting their functionality. This definition underscores an individual's capacity to engage in desired activities.

The brain operates on at least three primary levels influencing all facets of our daily lives: managing sensory interpretations and movement, overseeing emotional, mental, and cognitive processes, and upholding social cognition and standard behavior. Thus, brain health can be defined as the maintenance of the brain's peak integrity and mental capabilities relevant to one's age, regardless of the presence of any overt brain disorders. In more specific terms, brain health pertains to the brain's functionality across various domains, including social-emotional, cognitive, sensory, motor, and behavioral, enabling an individual to achieve their maximum potential throughout their life, irrespective of having a disorder or not.

Various factors influence brain health throughout one's life. These include safety and security measures, physical well-being, environmental factors, social connections, continuous learning, and access to quality services. Addressing these elements is essential for optimizing brain health. Doing so can enhance your physical and mental health and improve your social and economic standing, leading to overall better well-being.

Certain events and conditions can disrupt brain development and damage its structure or function over time. This includes neurological disorders, neurodevelopmental conditions, and congenital anomalies. Comprehensive brain care should be an individualized approach encompassing rehabilitation, treatment, promotion, prevention, and care. When addressing brain health, one must also consider personal experiences, professional endeavors, and familial or social relationships. Medically speaking, the neurological disorders that may affect brain function can be classified into:

- Brain conditions like brain tumors, traumatic brain injury, meningitis, and cerebrovascular diseases directly harm the brain's structure.
- Functional brain disorders, such as neurodegenerative diseases like Alzheimer's or mental conditions like bipolar disorder, impact brain networks.
- Some disorders, such as sleep disorders and migraines, do not show noticeable brain structure or function changes.

These neurological disorders can vary in their impact on brain function and health. Mood disorders, for instance, can affect emotional regulation and the processing of rewards. Alzheimer's disease, a form of dementia, leads to a deterioration in cognitive abilities. Issues like physical disabilities, balance disturbances, and aphasia can contribute to cognitive deficits if not addressed adequately by a doctor.

As we age, brain health becomes increasingly vital. It's commonly believed that brain function diminishes over time. In 2015, the global population of individuals over 60 years old was 900 million, and this number is projected to rise to two billion by 2050. As the population ages, there will likely be an increase in neurological, mental, and cognitive disorders, placing a significant economic strain on societies. In 2016, neurological disorders ranked as the primary cause of global disability and the second leading cause of death. Key contributors to neurological disabilities, as identified by the Global Burden of Diseases study, include conditions like meningitis, Alzheimer's, migraines, and strokes. Surprisingly, about one in four adults is at risk of experiencing a stroke after age 25. Given these statistics, it's crucial to ask: How healthy is your brain, and what steps are you taking to ensure its continued well-being?

Brain health is fundamental for problem-solving, decision-making, communication, and leading a fulfilling life. The brain is central to our daily functioning, arguably making it the most critical organ in our body. Factors influencing brain health include exercise, environment, diet, medical care, and sleep. Regular physical activity can counteract age-related brain decline, help control blood pressure, reduce the risk of vascular diseases that can lead to stroke, and ensure a consistent blood flow to the brain. This activity can range from sports and aerobics to simply taking a 15-minute walk. Mental exercises that bolster logical and strategic thinking also contribute to a robust brain health regimen.

Sleep is crucial for brain health. On average, humans sleep for about a third of their lives, benefiting the immune system, brain function, and overall well-being. Sleep enhances cognitive sharpness, decision-making, and concentration. Although individual sleep needs vary based on age, adults are generally advised to get between six and seven hours of sleep nightly to maintain optimal brain health.

Your environment and diet significantly impact brain health. Environmental contaminants in the air, food, and water can negatively affect brain function, potentially leading to neurodegenerative or neurodevelopmental disorders. For instance, repetitive head traumas, such as those experienced in some sports, can damage brain cells. Furthermore, a poor diet can exacerbate these adverse effects. Just as a balanced diet benefits the body, it also supports brain health. A diet rich in vegetables, whole grains, fruits, and proteins is essential for nourishing the body and brain.

Finally, brain health is linked with access to care and exposure to prevention programs. Access to care allows you to catch issues early and get the proper treatment to prevent escalation. This means that a lack of healthcare negatively impacts brain health. Caring for your brain protects you from disorders and diseases like:

- Mental health disorders
- Brain tissue injuries
- Degenerative diseases
- Vascular diseases
- Inflammation

Brain health is often assessed using the 'cognitive clock,' a tool researchers at Rush University Medical Center developed. This tool evaluates brain health based on an individual's performance on cognitive tasks. It aims to identify those at risk of cognitive and memory problems. The researchers believe this tool can help predict the risk of conditions like Alzheimer's. Although age is a primary risk factor for Alzheimer's, it doesn't guarantee the onset of the disease, as not everyone develops dementia as they age.

The cognitive clock evaluates how an individual's cognitive function compares to the average for their age group. While cognitive decline is natural with age, the rate varies among individuals. By analyzing data from long-term studies of diverse populations in Chicago, researchers established a baseline profile for cognitive aging. This clock represents the typical cognitive decline pattern based on age and provides a reference point. As a result, it's possible to assess whether an individual's cognitive function is above, below, or on par with their age group's average.

The development of the cognitive clock was based on data from 1,057 participants who underwent annual assessments for 24 years. These assessments included a mental state exam, which evaluates cognition by examining language, memory, attention, visual-spatial skills, and orientation. In addition to these exams, participants underwent comprehensive evaluations encompassing neurological exams, medical histories, and neurocognitive tests. These evaluations aimed to observe how cognitive abilities changed over time and establish a standard profile.

Researchers observed that, on average, cognitive age remains stable until around 80 years of age. There's a moderate decline from 80 to 90 years; after 90, the decline accelerates until death. The researchers tested the cognitive clock on a separate group of 2,592 participants to validate their findings. The results indicated that cognitive age is a reliable indicator of brain health. The cognitive clock offers a straightforward and accurate metric for evaluating cognitive performance.

The cognitive clock is a valuable tool, but a single test cannot capture its comprehensive health, function, and connectivity due to the brain's complexity. Clinicians often evaluate brain activity to understand its functionality, diagnose psychiatric conditions like depression or schizophrenia, or detect neurodegenerative diseases. Historically, tools like functional Magnetic Resonance Imaging (fMRI) and Electroencephalography (EEG) have been standard for monitoring brain activity. However, the field is advancing, with emerging companies investigating alternative techniques, such as measuring the brain's subtle magnetic fields. Here are some crucial factors that are used to shed light on brain health and function:

- **Amino acids, neurotransmitters, and hormones:** Neurotransmitters and hormones play essential roles in neural communication and body signaling. Disturbances in their functioning can indicate potential issues. Amino acids, as the precursors of hormones and peptides, offer valuable insights when assessing their levels concerning brain health.

- **Electrical signaling:** Neurons convey messages using electrical signals initiated by ion movement across their membranes. The EEG tool captures this electrical activity, providing a perspective on brain health and highlighting potential neuronal disorders.
- **BDNF levels:** The Brain-derived neurotrophic factor (BDNF) facilitates neurogenesis and is instrumental in neuron growth and differentiation. It plays a significant role in memory and learning processes. Reduced BDNF levels are associated with diseases such as Alzheimer's.
- **Neuronal firing patterns:** Although neurons communicate primarily through electrical signals, different brain areas showcase specific signaling patterns. Analyzing these patterns can assist in identifying potential neural challenges.
- **Cognitive function:** Key indicators of brain health include capabilities such as short-term and long-term memory, concentration, alertness, and adaptability. Continuous

cognitive impairments can suggest a deeper underlying issue.

Understanding the multifaceted aspects of brain health and function requires a comprehensive approach considering various factors. Recognizing these key indicators opens doors to optimizing brain wellness and addressing underlying challenges impacting our cognitive abilities.

PEPTIDES: A COGNITIVE ENHANCEMENT TOOL

Much like a computer, the brain is constantly evolving and seeking peak performance. Just as computers rely on software updates to fix glitches and improve operations, the brain utilizes neuropeptides to optimize communication between its cells. These neuropeptides are vital in regulating various functions, including appetite, sleep, and mood modulation.

In this computer-like system of the brain, different regions can be likened to various software applications or programs. Neuropeptides act as the updates or patches, bridging these programs for synchronized function. Depending on the specific neuropeptide, it might optimize the system for restful sleep or ramp up performance for alertness. Just as a software update can introduce new features to a computer or improve its speed, different neuropeptides can fine-tune our emotions and actions. Some might enhance mood, while others could act as the body's natural antivirus, combating pain.

People, much like computers, may react differently to certain 'updates' or neuropeptides. The effects of specific neuropeptides can vary from person to person, impacting responses to stress or interactions with substances like alcohol. Neuropeptides communicate commands to particular molecules and cells in the body. In peptide therapy, specific peptides target the brain cells associated with cognition. Their goal? To stimulate cell growth, bolster cognitive function, and foster brain flexibility.

Classic brain peptides are grouped into several families, such as neurohypophyseal hormones (like vasopressin and oxytocin), hypothalamic-releasing hormones (like TRH and CRH), pituitary peptides (like growth hormone and ACTH), opioids (like endorphins), tachykinins (like substance P), and peptides also found in the gastrointestinal system (like cholecystokinin). These peptides are distributed throughout the central nervous system. Notably, anterior brain regions such as the cerebral cortex, striatum, and amygdala contain a broader range of neuropeptides than posterior regions like the cerebellum and spinal cord. Even within areas that share a developmental origin, such as the thalamus and hypothalamus from the diencephalon, the neuropeptide compositions can vary significantly.

Our understanding of how brain peptides regulate eating and body weight has evolved considerably. In the 1950s, researchers identified a genetic mutation that, when recessive, causes severe obesity and type II diabetes. This discovery suggested that the protein produced by this gene might be involved in regulating fat storage in the body. Mice with mutations in this gene become overweight, display reduced metabolic rates, activity, and body temperature, and develop diabetes. Subsequent research revealed that the gene produces a hormone called leptin. Primarily made in fat tissue, leptin plays a key role in a feedback system that manages body weight by influencing food intake and energy expenditure. This regulation occurs in the hypothalamus through the action of certain brain peptides, one of which is neuropeptide Y, which is closely related to weight control.

The brain utilizes particular neuropeptides in its feedback systems. One such group of neuropeptides is the orexins, also termed hypocretins, due to their similarity to a gut hormone named secretin. Two primary types of orexins are orexin-A and orexin-B, composed of 33 and 28 amino acids, respectively. These orexins are produced by neurons in the hypothalamus, specifically within the perifornical nucleus. These neurons derive orexins from a larger protein known as the preproorexin precursor. Once made, orexins are dispatched to various brain regions, including the olfactory bulb, cerebral cortex, thalamus, hypothalamus, and brain stem. Within these areas, orexins bind to specific receptors, modulating the release of neurotransmitters and thereby influencing neuronal communication.

Orexin neuropeptides are involved in sleep regulation. Studies in mice lacking the orexin gene have demonstrated abnormal sleep patterns. Similarly, specific gene mutations in Doberman Pinschers lead to sleep disturbances reminiscent of narcolepsy observed in humans. Moreover, many individuals diagnosed with narcolepsy exhibit decreased orexin levels, further underscoring the relationship between orexins and sleep disorders.

Opioid peptides modulate the brain's activity, altering the function of other neurotransmitters in the central nervous system. They influence how nerve cells respond to electrical signals, which can regulate the release of various neurotransmitters. These modifications can result in effects such as pain relief and euphoria. Furthermore, opioid peptides play a role in certain behaviors, including alcohol consumption. Activation of the opioid system by alcohol may reinforce its use and lead to excessive drinking. Research indicates that inhibiting the opioid system decreases alcohol intake in animals and reduces cravings and consumption in human alcoholics. Some individuals may possess a genetic predisposition, making them more sensitive to alcohol's effects on the opioid system, potentially increasing their susceptibility to alcoholism.

Research into the use of peptides for treating neurological disorders is ongoing, with promising results emerging. For instance, MIT researchers have identified a potential treatment for Alzheimer's disease. They targeted an enzyme, CDK5, found in excessive activity in the brains of those with the disease. After treating mice with a specific peptide designed to counteract this overactive enzyme, researchers observed notable decreases in brain and DNA damage. Furthermore, these treated mice demonstrated enhanced learning abilities, as evidenced by their improved performance in tasks such as navigating a water maze.

The MIT researchers observed that the peptide treatment decreased brain damage and inflammation while enhancing mice's behavioral performance. While more studies are required, there's hope that this peptide might be a potential treatment for Alzheimer's disease and other dementias associated with an overactive

CDK5. A significant benefit of this peptide is its specificity; it doesn't affect CDK1, an enzyme structurally similar to CDK5 but essential for various functions. Additionally, the peptide's size is comparable to other peptide drugs used in medicine.

CDK5 is an enzyme critical for central nervous system development and synaptic regulation. In conditions like Alzheimer's disease, CDK5 becomes overly active due to its interaction with a smaller protein called P25. This heightened activity causes CDK5 to impact molecules it typically wouldn't, leading to the development of neurofibrillary tangles, a key characteristic of Alzheimer's disease. Previous attempts to target P25 had undesirable side effects, primarily from unintentional interactions with other enzymes. Addressing this, the MIT researchers developed a peptide composed of 12 amino acids designed to interact with CDK5 and mitigate the problem. This innovation underscores the potential of peptides in advancing brain health.

ACHIEVING SUPERIOR BRAIN HEALTH AND COGNITION

Recently, the use of peptides to improve cognition has gained traction due to their effectiveness. Peptides are functional and typically exhibit minimal side effects.

They have been reported to enhance mood, mental clarity, focus, and overall productivity. Here is a list of some common peptides to improve brain function:

- **Selank:** This peptide interacts with GABA neurotransmitters, making it a potential treatment for generalized anxiety disorder. Administered as a nasal spray or orally, its onset of action takes a few days. Its mechanism involves an increase in BDNF levels.
- **DIHEXA:** Showcasing potential for treating conditions like Parkinson's and Alzheimer's, DIHEXA functions by repairing brain synapses, enhancing receptor sensitivity, and modulating gene expression. This leads to improved brain function and clarity.
- **RG3:** Primarily administered as a nasal spray, RG3 combats neurotoxicity resulting from inflammation, a common cause of brain fog. It enhances memory, attention, and focus. When paired with B12 and NAD+, its benefits extend to improved energy and productivity. RG3 promotes mitochondrial function, mimicking cellular effects seen post-exercise.
- **Noopept:** Beneficial for preventing cognitive impairment, noopept stimulates BDNF and

protects against oxidative brain damage. It's recommended for enhancing spatial memory.

- **Semax:** Ideal for those seeking to enhance learning and memory, Semax elevates BDNF levels, providing potential relief from depressive disorders and anxiety. Its mechanism involves gene modulation and immune response tempering. While effective on its own, combining it with peptides like DIHEXA can optimize its benefits.
- **Cerebrolysin:** This peptide promotes nerve repair and regeneration. It can cross the blood-brain barrier, directly influencing neurons. Its effects include enhanced mental clarity, motivation, and reduced fatigue. Increasing BDNF levels and decreasing inflammation offers potential benefits for conditions like stroke and traumatic brain injuries.
- **FGL:** Influencing neurotransmitter hormone levels, like testosterone, FGL operates similarly to neural cell adhesion molecules. This leads to the formation of mushroom dendrites associated with improved long-term memory. FGL can enhance memory and cognitive performance and may also alleviate depressive symptoms.

Selecting the right peptides for brain health requires careful consideration based on individual needs. While the combinations mentioned above can serve as starting points, the most suitable choices will vary depending on specific conditions (detailed further in Chapter 5). For instance, combining Cerebrolysin and DIHEXA might be beneficial if addressing autoimmune issues and fatigue. A combination of RG3 and nootropics, such as oxytocin, could be considered for enhanced focus. For conditions like Alzheimer's or MS, starting with DIHEXA, Semax, and Cerebrolysin might be advisable. Always seek a physician's advice before beginning any treatment.

INTEGRATING PEPTIDES WITH OTHER BRAIN-BOOSTING STRATEGIES

After delving into the intricacies of peptides and their profound effects on brain function, it's essential to understand that the brain, a marvel of complexity, can be influenced by many factors. While peptides represent a groundbreaking approach to neural enhancement, they are just one piece of the puzzle. Combining peptide treatments with other well-researched strategies can create a synergistic effect, magnifying the benefits to our cognitive health. In this section, we'll explore these complementary methods that, when combined with peptides, present a comprehensive approach to nurturing and optimizing our brain's potential.

Beyond the realm of peptides, a myriad of other strategies exist, each offering unique pathways to optimal cognitive health. Enter the world of "neurohacking." This approach marries scientific research with practical interventions, offering a holistic toolbox to enhance brain function. Whether it's exercise, diet, or cutting-edge technologies, each avenue offers distinct benefits and provides a well-rounded strategy for brain enhancement.

High-intensity interval training (HIIT) is a recommended form of exercise with proven health and neurological benefits. Studies indicate that HIIT stimulates the production of Brain-Derived Neurotrophic Factor (BDNF). Elevated BDNF levels have been associated with improved cognitive function in various age groups, from children and adolescents to older adults, as well as in individuals with conditions like multiple sclerosis.

Neurostimulation is a technique that uses electric currents for therapeutic purposes, often to treat chronic pain. Some evidence suggests that it enhances cognitive abilities. For instance, epilepsy patients and individuals recovering from addiction have reported cognitive benefits from neurostimulation. A specific study in the NeuroImage journal indicated that neurostimulation might increase brain processing speed and support learning.

The ketogenic (keto) diet is a low-carb, high-fat diet designed to shift the body's metabolism to burn fats for energy, producing ketones. Originally used to treat epilepsy, the keto diet has been studied for its potential benefits on brain health. Research suggests that the diet may improve cognitive performance, enhance working memory, and provide some benefits for neurodegenerative conditions.

Moderating alcohol intake is essential for brain health. Heavy alcohol consumption can detrimentally affect brain functions and cognitive performance. The effects of moderate alcohol consumption are complex, with studies producing varied results. Some research suggests potential adverse effects at even moderate levels, while other studies have found that moderate alcohol consumption may correlate with better cognitive performance than complete abstinence.

Consuming sufficient antioxidants is essential for combating harmful free radicals produced during metabolic processes. Free radicals can harm DNA and cell structures. Maintaining mitochondrial function, crucial for cellular energy production, depends on neutralizing these radicals. Natural sources of antioxidants include vegetables, fruits, and teas. Some supplements, such as resveratrol, pterostilbene, and EGCG, have shown potential in mouse studies to support memory and cognitive functions.

Finally, diet and sleep are critical factors for brain health. A well-balanced diet provides essential nutrients that support brain function. While there are various diets, many foods benefit the brain. Foods rich in Vitamin D promote neural growth, synaptic connections, and neurotransmitter activity. B vitamins offer protection against neurotoxins and support neuron communication. Magnesium threonate can penetrate the blood-brain barrier and mitigate oxidative stress in neurons. Ingredients like zinc and NAC are fundamental for hormone production and cell protection. Additionally, zinc, ALA, curcumin, and fish oil possess anti-inflammatory properties beneficial for the brain.

Pairing a balanced diet with adequate sleep is essential for optimal cognitive health. Managing sleep disturbances, especially those caused by stress, is crucial. Implementing methods such as meditation and ensuring a sleep-friendly environment can be beneficial. By adopting these practices, you set the stage for sustained cognitive well-being. Adding peptides to this regimen can further enhance cognitive performance.

Understanding the potential of peptides for brain health is crucial. How will this knowledge shape your decisions? Imagine the impact of enhanced memory and focus on your daily activities. What actions will you take to maximize the advantages of peptides?

Peptides can significantly improve cognitive clarity and focus, setting a distinct contrast against a distracted and hazy mind. However, the scope of peptides extends beyond cognition. In the upcoming chapter, we will explore how peptides influence aging, longevity, and even aspects of our intimate lives.

4

ENHANCING LONGEVITY AND SEXUAL VITALITY WITH PEPTIDES

Is it possible to age gracefully, maintaining health and vitality throughout our years? Could scientific advancements help sustain our physical strength even as time advances? As aging concerns many, there's a growing interest in understanding the mechanisms that influence our longevity and quality of life. Research into this area has spotlighted the potential role of peptides. This chapter delves into the science behind peptides and their possible impact on aging, longevity, and sexual well-being. Peptides may offer ways to prolong our health and enhance our overall life experience.

Aging brings with it a series of cellular and molecular changes, from DNA damage to reduced cellular repair mechanisms. This chapter delves into these age-related processes, offering a comprehensive understanding of how they influence our health and longevity. We will discuss recent research that points to the potential benefits of peptides in counteracting some of these age-driven declines. Topics will include the potential role of peptides in strengthening immunity and promoting cellular rejuvenation. Furthermore, we'll explore the relationship between peptides and sexual health, providing insights into how these compounds might enhance our overall quality of life.

THE SCIENCE OF AGING

Cells are the basic building blocks of all living tissue. Different types of cells have distinct structures and functions, yet they all share common features. When similar cells collaborate to fulfill a specific function, they form tissues. These tissues then combine to create organs. There are four primary tissue types:

- **Epithelial tissue:** This tissue forms the covering of both external and internal surfaces of the body, such as the skin.
- **Connective tissue:** This provides support and binds other tissues together. Examples include bone, blood, and skin-supporting tissues.

- **Nervous tissue:** Comprising neurons or nerve cells, this tissue transmits signals throughout the body. The brain and spinal cord are composed of nerve tissue.
- **Muscle tissue:** This encompasses skeletal muscles that move bones, smooth muscles found in organs like the stomach, and cardiac muscle that makes up the heart.

With age, every tissue type and the organs they form experience a decline in function. Cellular changes are universal as we age. Cells find it harder to divide and grow, increase in size, and often accumulate more lipids (fats). Over time, many cells either lose their functionality or don't work as efficiently. Waste products begin to build up within tissues. Connective tissues become less flexible, increasing rigidity in blood vessels, airways, and various organs. Changes in cell membranes make it challenging for tissues to receive necessary nutrients and oxygen and to expel waste products, including carbon dioxide. Many tissues undergo atrophy, meaning they reduce in size, and some may become irregular or hardened.

As cells and tissues age, organ functionality also declines. However, the decline in organ function is gradual for most people because most individuals don't often push their organs to maximum capacity. Organs have a built-in reserve capacity, allowing them to operate beyond normal requirements. For example, the heart of a 20-year-old can pump approximately ten times the blood required for essential bodily functions. However, after age 30, this reserve decreases by about 1% annually. The most significant reductions in organ reserve are observed in the kidneys, lungs, and heart. The extent of reserve loss varies from person to person and across different organs.

As you age, there are gradual changes in organ functionality. When an organ is pushed beyond its usual capacity, it can't compensate by increasing its function, which can lead to issues like sudden heart failure. Factors that stress the body, such as illnesses, significant life changes, certain medications, or increased physical demands, can exacerbate these issues by pushing the body to its limits. Furthermore, a diminished organ reserve makes it challenging for the body to maintain balance. For instance, the liver and kidneys process drugs, but as their reserves decline with age, they do so less efficiently. This reduced efficiency can necessitate lower drug dosages and lead to increased side effects in older individuals.

Additionally, full recovery from illnesses becomes less certain, resulting in accumulating health issues. As people age, they might also experience drug side effects that mimic disease symptoms, leading to potential misdiagnoses. Moreover, some medications may affect older individuals differently than younger ones.

The reasons behind aging remain largely elusive, and there isn't a single explanation for why people age. Some theories suggest that long-term exposure to ultraviolet light, metabolic byproducts, or general wear and tear in the body contribute to aging. Others posit that aging is a preordained process controlled by our genes. However, no single theory can account for all the variations and complexities of aging. Aging is an intricate process affecting individuals and their organs in diverse ways. Gerontologists, experts in aging, believe that aging results from multiple factors throughout one's life. These include cultural influences, environmental exposures, genetic makeup, leisure activities, physical exercise, past illnesses, and dietary habits, to name a few.

Aging differs significantly from the predictable transformations experienced during adolescence. Each individual ages at their own pace. While some bodily systems may show signs of aging as early as 30, others remain unaffected until later in life. The specific rate and pattern of aging are unique to each person. However, as aging progresses, certain cellular changes can be anticipated:

- **Atrophy:** This refers to the reduction in cell size. When many cells undergo atrophy, an entire organ can shrink. This phenomenon can occur in any tissue but is most prevalent in the sex organs, brain, heart, and skeletal muscles. Causes of atrophy include decreased workload, reduced activity, inadequate cell nutrition, diminished blood supply, and lack of nerve stimulation. Consequently, bones may become more fragile and susceptible to fractures.
- **Hypertrophy:** This is the enlargement of cells due to an increase in the protein content within cell structures and membranes. As aging progresses and some cells shrink, others might enlarge to compensate for the loss of cell mass.
- **Hyperplasia:** Hyperplasia is characterized by an increase in the number of cells resulting from an uptick in cell division. This process compensates for cell loss and aids in tissue and organ regeneration, especially evident in the skin, liver, and bone marrow. However, some tissues, like cartilage and smooth muscles, have limited regenerative capabilities, while others, such as heart muscles and nerves, don't regenerate at all.
- **Neoplasia:** This term describes the formation of tumors, which can be benign (non-

cancerous) or malignant (cancerous). Neoplastic cells tend to have irregular shapes and functions and proliferate rapidly.
- **Dysplasia:** Dysplasia involves changes in the organization, size, or shape of cells, leading to abnormalities. This is commonly observed in the lining of the respiratory tract.

PEPTIDES AND CELLULAR REGENERATION

As you age, various physiological changes occur throughout your body. There's a shift in hormone production, a potential decline in immune function, and visible alterations like wrinkles and looser skin. Sleep patterns might change, and you may experience discomfort in your joints and muscles. Additionally, changes in the appearance of your face and alterations in the function of your reproductive system can occur, mirroring changes in the kidneys, lungs, and nervous system. However, these age-related changes aren't inevitable. Peptides offer a potential approach to mitigate some of these aging effects at the cellular level.

Peptides are crucial for cellular regeneration, acting as messengers coordinating various cellular activities. They function as signaling molecules, facilitating communication between cells to promote growth, repair, and renewal. By interacting with specific cell receptors, peptides trigger a series of internal events affecting gene expression, protein production, and cell division. This exact regulation by peptides is vital for tissue repair and sustaining proper organ function.

Bone tissue engineering has seen significant advancements, especially with the inclusion of peptide-related research. Several peptides that aid bone healing have been identified, positioning them as potential therapeutic agents. Experiments, both in vitro and in vivo, have consistently demonstrated the efficacy of these peptides in enhancing bone regeneration, suggesting their promising potential for medical use.

When bones experience trauma, the body's musculoskeletal system initiates a repair response. However, this natural healing mechanism doesn't always prove effective. About 5% to 10% of fractures don't heal correctly. Sometimes, the bone's healing is delayed, referred to as "delayed union." In other instances, the bone doesn't heal at all, a condition called "non-union." Non-union fractures can be particularly problematic for patients, requiring prolonged treatments and frequently leading to significant pain and discomfort.

Peptides have the potential to expedite bone healing. They function by attaching to specific receptors on target cells, thereby modifying the cells' activities. Some peptides are crafted to boost the creation of bone tissue. While peptides originating from bone morphogenetic proteins (BMP) are extensively researched, other peptides also exhibit potential. For instance, a peptide consisting of 34 amino acids derived from parathyroid hormone (PTH) has been identified to promote the development and differentiation of osteoblasts, the cells responsible for bone formation.

Research on rabbits has shown that using a synthetic matrix infused with the peptide derived from PTH can notably boost bone regeneration in fractured regions. In animal tests, consistent subcutaneous administrations of this peptide resulted in heightened bone mineral levels, greater bone density, and increased overall bone mass, coupled with enhanced bone strength. Clinical observations have also indicated successful results when employing this method for bone complications, such as fractures that have difficulty healing.

A study on post-menopausal women with distal radial fractures explored the outcomes of daily injections with a synthesized version of the PTH peptide. While the higher dosage didn't produce a notable difference compared to the placebo, the lower dosage resulted in quicker healing. Additionally, observational evaluations indicated that those treated with the PTH-derived peptide displayed enhanced callus development. The ongoing research into peptides and their role in bone healing suggests promising avenues for advancing the management of bone injuries, potentially expediting recovery and enhancing patient results.

Peptides have demonstrated promise in promoting skin tissue regeneration. Researchers, through genomics, identified a specific peptide derived from the skin of the Asian frog species, *Odorrana andersonii*. This peptide exhibited wound-healing properties. Laboratory experiments revealed that when this peptide was applied to keratinocytes (skin cells), it enhanced their repair speed and decreased the release of specific inflammatory compounds. Furthermore, in a mice wound healing experiment, wounds treated with a solution containing this peptide healed more rapidly.

Research has led to the identification of peptides with varied functionalities. A peptide derived from the keratin breakdown of feathers showed antioxidant properties. Another peptide promoted the growth of specific cells related to our teeth and influenced them to take on characteristics that help form mineral-like structures. Additionally, particular peptides have shown the ability to inhibit angiogenesis—the process by which new blood vessels form—and, as a result, slow down tumor growth. Blocking angiogenesis can be vital because tumors rely on these new blood vessels to receive nutrients and grow.

Peptides assist in directing circulating cells to specific sites in the body, supporting tissue repair. They also influence stem cell differentiation. Over time, stem cells can degrade, losing their efficiency in self-renewal and varying cellular transformations. This decline in stem cell functionality is closely associated with many age-related health concerns. Aging can compromise stem cell operations, potentially leading to cell death, diminished cell division, or impaired tissue regeneration. A practical illustration of this phenomenon is the extended healing time required for bone fractures in older people compared to younger individuals.

In regenerative medicine, peptides amplify our stem cells' activity, facilitating quicker healing and restoration of injured tissues. An example is the peptide PEDF, which has attributes that reduce inflammation and inhibit new blood vessel formation—a specific segment of PEDF aids in the proliferation and differentiation of neurons and diverse stem cells. A subset of this segment, termed PEDF-derived short peptide (PDSP), has shown potential in boosting the growth of specific stem cells, such as those found in the cornea and meibomian gland. Consequently, this can be beneficial in addressing dry eye disease by promoting corneal cell regeneration, diminishing inflammation, and aiding in the recovery of the meibomian gland.

PDSP has also demonstrated potential in stimulating the proliferation of mesenchymal stem cells, which can differentiate into various tissues such as bone, cartilage, muscle, fat cells, and connective tissue. Given this capability, PDSP holds promise for applications in conditions like osteoarthritis, where it could aid in cartilage regeneration, mend tissue damage, and alleviate joint discomfort.

Peptides have also carved a significant niche in the beauty industry due to their myriad benefits for skin health. With age, the decrease in collagen and elastin fibers makes the skin less resilient, leading to wrinkles and a loss of firmness. Peptides address this by boosting collagen levels and enhancing skin elasticity. Their role in skincare is further emphasized by their ability to stimulate skin repair and combat signs of aging. A 2020 study in the International Journal of Molecular Sciences highlighted the effectiveness of peptides in reversing signs of premature aging. Participants over the age of 40, upon applying peptides to their face and neck for two weeks, reported visible improvements in their skin's appearance. Furthermore, some peptides can mimic the effects of mild Botox, aiding in relaxing facial muscles and diminishing fine lines.

Beyond improving skin texture, peptides protect skin against environmental threats like UV rays, bacteria, and pollutants. Some peptides have shown potential in shielding the skin from UV damage, though more research is needed to fully realize their capability in sun protection. Additionally, peptides possess antimicrobial properties that target bacteria. This makes them especially beneficial for addressing skin conditions like acne, which often result from bacterial blockages in pores. By reducing bacterial buildup, peptides can help prevent pore blockages and the subsequent formation of acne.

Given these attributes, it's no wonder that peptides are frequently incorporated into cosmetic and skin care products, fostering skin cell renewal and offering a holistic approach to skin health. With many peptides available, selecting the most suitable ones for your skincare regimen can be difficult. Here are three of the most widely used peptides for skin care and anti-aging:

- **GHK-Cu:** This peptide elevates collagen synthesis, diminishes wrinkles, and enhances skin elasticity. Research indicates its capacity to mitigate inflammation, boost skin hydration, and accelerate the recovery process following tissue damage.

- **Matrixyl:** This peptide can invigorate collagen production, reduce the appearance of wrinkles, and enhance skin suppleness. Scientific research has uncovered its capability to increase levels of hyaluronic acid in the skin, contributing to optimal hydration and flexibility.
- **Argireline:** Argireline has demonstrated its efficacy in wrinkle reduction, skin tone enhancement, and collagen synthesis stimulation. It offers the added benefit of relaxing facial muscles, diminishing the visibility of dynamic wrinkles.

PEPTIDES FOR LONGEVITY AND AGE-DEFYING BENEFITS

Peptides have numerous anti-aging properties that may support a longer and healthier lifespan. They enhance cellular health, reduce inflammation, and strengthen the body's natural protective mechanisms, offering a comprehensive approach to longevity. With age, cells naturally deteriorate, leading to diminished function and energy. Peptides respond by boosting the synthesis of vital proteins that maintain cellular health. Inflammation, a primary factor in aging, is associated with several health complications like heart disease, diabetes, and cognitive impairments. Peptides help reduce inflammation, potentially preventing or delaying these age-associated health challenges.

Peptides not only impact cellular health and inflammation but also bolster the body's natural defenses against stress and disease. Some peptides boost the production of antioxidant enzymes, which help guard against oxidative stress and cellular damage. Others strengthen the immune system, providing enhanced protection against infections and related health issues. Incorporating peptide therapy can aid in maintaining optimal health and appearance as one ages.

A study conducted in Japan explored the effects of collagen peptides on aging-related skin issues among participants. The study tracked 66 Japanese women, all over 40, monitoring their skin for changes. For 56 days, half received 10g of collagen daily, while the other half were given a placebo. Results showed that the collagen group significantly increased skin moisture, as measured by bioelectric impedance analysis. In a similar study in France with women over 40, those receiving collagen for three months also exhibited a notable increase in skin moisture compared to those on a placebo.

IGNITING PASSION: PEPTIDES AND SEXUAL ENHANCEMENT

Peptides also have the potential to enhance libido and improve sexual well-being. Reduced sexual desire is a common challenge associated with aging. Factors such as demanding schedules, work stress, and other daily pressures can deplete energy and reduce interest in intimacy.

A decline in libido can have lasting effects on personal experiences and relationships. Peptides offer potential solutions for such concerns. Recent advancements in peptide therapy have highlighted its potential to improve sexual function. Specifically, bremelanotide (PT-141) has been shown to effectively address both female sexual dysfunction and male erectile dysfunction safely.

Female sexual dysfunction (FSD) includes various issues such as pain during intercourse, decreased libido, problems with lubrication, and reduced sexual satisfaction. Bremelanotide has shown the potential to improve arousal, lubrication, and the quality of orgasms in women.

In men, erectile dysfunction (ED) is a common issue, and its incidence increases with age—for example, about 20 percent of men over 60 experience moderate

ED. Bremelanotide has shown potential benefits in addressing this condition, enhancing erection quality, libido, and overall sexual performance.

Peptides that enhance libido often have additional benefits, such as elevated energy levels and muscle growth. Besides bremelanotide, sermorelin and melanotan II are other peptides that can boost sexual desire. Sermorelin, a 29 amino acid peptide, promotes libido by triggering the release of growth hormones through the pituitary gland. This action not only elevates sexual desire but also augments strength, endurance, and bone density. On the other hand, melanotan II increases libido and stimulates melanocytes, resulting in a tan without sun exposure. Additionally, it aids in cholesterol metabolism and suppresses appetite.

In this chapter, we've explored the potential of peptides to promote longevity and vitality. We've delved into the aging process and examined how various peptides can mitigate its effects. Peptides are vital for cellular communication, affecting many functions, from tissue repair to immune responses. They offer protection against age-related conditions and bolster overall health. Additionally, peptides can improve sexual health by addressing concerns such as erectile dysfunction and hormonal imbalances, potentially rejuvenating intimate relationships.

The question is, will you put this information into action for a healthier and more vibrant life? Consider consulting with a medical professional to explore peptide therapy as a potential addition to your wellness regimen. You could unlock a new level of ageless strength and vitality with peptides. In the upcoming chapter, we'll guide you on effectively incorporating peptide therapy into your daily routine, providing a clear starting point.

5

MAKING PEPTIDE THERAPY WORK FOR YOU

As of 2023, more than a million people around the globe have turned to peptide therapy as a groundbreaking avenue for health enhancement. Are you contemplating joining their ranks? If so, this chapter is your comprehensive guide to embark on this transformative journey, equipped with the essential knowledge you'll need for a safe and effective experience.

Peptides have gained recognition as a revolutionary contribution to the medical field, providing pivotal roles in many biological functions. Their therapeutic applications are flourishing at an unprecedented rate, offering potential solutions to various health-related issues—from combating signs of aging to enhancing sexual health and beyond. However, it's crucial to approach this potent form of therapy with due diligence, understanding its incredible promise and inherent risks.

This chapter aims to give you the information needed to make well-informed decisions about peptide therapy. We'll delve into its diverse benefits, which have already significantly impacted many people's lives. Additionally, we will discuss the risks and possible side effects to be aware of, ensuring you are prepared to engage with this therapy safely.

Ethical considerations also play a significant role in the world of peptide therapy. As the field rapidly evolves, it poses questions about equitable access, long-term impacts, and the ethical use of such substances. We will provide insights into these concerns so you can make choices that align with your ethical principles.

By the end of this chapter, you'll be better equipped to embark on your peptide therapy journey conscientiously and informed. The information presented here aims to empower you, allowing you to take control of your health in a new and innovative way, fully aware of the responsibilities that come along with it.

WHY YOU SHOULD CONSIDER PEPTIDE THERAPY

Peptides occur naturally within the human body. These crucial biological molecules are involved in a myriad of functions, from modulating the immune system to regulating metabolism. However, the body doesn't always produce peptides in the quantities needed for optimal health, especially as we age. This is where the increasingly popular and innovative approach of peptide therapy steps in, designed to address a broad spectrum of health conditions and improve overall bodily performance.

The administration of peptide therapy can take multiple forms. These range from precise dose injections to capsules, topical creams, and even inhalation methods. These therapies are generally considered natural and non-invasive, and they offer a multitude of potential benefits. Some promising effects include reducing inflammation, accelerating healing, catalyzing weight loss, stimulating hair growth, and much more. These therapies aim to combat the inevitable physiological declines associated with aging.

Peptides are incredibly versatile, with over 7,000 known varieties playing roles in various biological functions. They can act as neurotransmitters in the brain and hormones in the bloodstream. For instance, some peptides facilitate communication between nerve cells, initiating complex processes throughout the body. Once released into the bloodstream, others act more like hormones, binding to specific cell receptors and launching intricate signaling pathways that govern a host of physiological processes.

Moreover, peptides serve as precise molecular messengers, issuing targeted instructions to cells. Some peptides, such as growth hormone-releasing peptides (GHRP), mimic the actions of human growth hormones. Others are formulated to focus on specific outcomes, such as weight loss or alleviating joint and muscle pain. They can act as guardians of metabolic processes, muscle mass, sexual vitality, and immune system health, among other roles.

The production of peptides within the human body is a dynamic process that can vary from person to person and diminish with age. Peptide therapy employs synthetic peptides as functional surrogates to compensate for this natural decline. The aim is to restore a state of balance, offering a range of benefits, including but not limited to muscle-sculpting, weight loss, energy boosting, physical performance improvement, sleep quality enhancement, libido increase, cognitive acuity sharpening, and even nerve and cartilage repair.

When natural peptide levels drop, individuals may experience various health problems, such as hormone imbalances, digestive issues, weight gain, chronic inflammation, and diminished libido. A decline in peptide production may manifest in symptoms like excess fat around the midsection, decreased lean body mass, mood disorders like anxiety and depression, fatigue, decreased bone density, lower exercise tolerance, lack of energy, dull skin and hair, and joint and muscle pain.

Peptide therapy addresses these issues by supplementing or adjusting the levels of these critical molecules in your body, thereby improving your health. These therapeutic peptides may be sourced from various animals or plants, including milk, eggs, meat, soy, hemp seed, oats, flaxseed, and wheat.

Its adaptability and versatility set peptide therapy apart, making it a personalized approach to healthcare. Specific peptides are selected based on their proven ability to activate particular biological functions. Though peptide therapy is not a quick fix—it typically includes a 'loading' phase of 3 to 6 months before full effects are realized—initial improvements can often be noted within a few weeks of treatment. And since peptides are naturally occurring substances, they are generally well-tolerated with minimal side effects.

WHAT TO EXPECT WHEN STARTING PEPTIDE THERAPY

Initiating a peptide therapy regimen often involves a carefully orchestrated daily routine tailored to the individual's health goals and profile. While the most prevalent method of peptide administration is via injections, oral capsules are also available as an alternative. Sometimes, a combination of injection and oral administration may be recommended to achieve specific therapeutic objectives. If you're considering insulin peptide therapy as an example, here's a general outline of what the process could look like for you.

The journey toward insulin peptide therapy often starts with an initial evaluation, including a fasting blood glucose test or a Hemoglobin A1c test. Conducted at a local clinic, these assessments gauge your body's ability to regulate sugar levels. The tests are a preliminary step to determine your suitability for insulin peptide therapy. If you meet the initial criteria, your healthcare provider will move forward with a more comprehensive set of tests and assessments. These evaluations will be tailored to your specific health profile and wellness goals, ensuring you're an appropriate candidate for the therapy.

Once your healthcare provider confirms that you're a suitable candidate for insulin peptide therapy, you'll move on to the actual treatment. Your healthcare professional will begin by sanitizing the selected injection site, often chosen in the upper hip or gluteal area. After ensuring the area is clean, the injection is administered according to your personalized treatment plan. You may require one or two daily injections depending on your specific needs and your healthcare provider's recommendations. It's not uncommon for insulin peptide therapy to be administered in two separate doses: a milder dose in the morning and a more substantial one before sleep.

It's important to note that while the benefits of peptide therapy usually start to manifest around the three-week mark (approximately 21 days into the treatment), results can vary from person to person. Don't be disheartened if you don't observe significant improvements within this initial period. Peptide therapy is generally a gradual process, and it may require a commitment of three to six months of consistent injections to start experiencing the full range of benefits. This is a nuanced and individualized therapeutic journey that rewards patience and consistency.

In summary, peptide therapy is a highly personalized medical treatment that should be embarked upon with appropriate medical supervision. It can offer various health benefits but demands precise administration and patient commitment for optimal results like any medical intervention.

RISKS AND SIDE EFFECTS OF PEPTIDE THERAPY

Peptide therapy is increasingly recognized as a safe and effective medical intervention for various health issues. Backed by the FDA's approval of numerous peptides for medical applications and ongoing clinical trials, our understanding of the benefits and potential risks of peptide therapy is continually expanding.

While most individuals tolerate peptide therapy well, minor side effects can be increased hunger or thirst, mild drowsiness, dry mouth, or slight itching at the injection site. These side effects tend to be dose-dependent and are generally classified into several categories:

- **Allergic reactions:** Though rare, some people may experience allergies, such as difficulty breathing, swelling, or hives.
- **Cardiovascular effects:** Certain peptides that impact heart function could lead to conditions like tachycardia, hypertension, or palpitations.

- **Cognitive effects:** When peptides interact with the central nervous system, you might experience fatigue, headaches, or dizziness.
- **Gastrointestinal effects:** Some peptides can affect the digestive system, leading to symptoms like diarrhea, nausea, or vomiting.
- **Drug interactions:** Although uncommon, peptides could interact with other medications. Therefore, it's crucial to disclose all the medications you're currently taking to your healthcare provider.
- **Hormonal imbalances:** If not properly administered, peptides that interact with hormones, like growth hormones, could cause imbalances.

Regarding peptides designed for skin care, preliminary research suggests they are generally safe and free from side effects, primarily because their small molecular size allows them to penetrate the skin deeply without causing surface irritation. However, concerns have been raised about whether these small molecules can enter the bloodstream and what long-term effects they might have. If you have sensitive skin or experience irritation or inflammation while using peptide-based skincare products, it's advisable to consult a dermatologist to discuss your options.

Since peptide therapy is relatively new, its long-term effects are not yet fully known. Therefore, working closely with a qualified healthcare provider is vital to determine an effective and safe treatment regimen tailored to your needs. Remember that peptide therapy should complement, not replace, a healthy lifestyle. Exercise, a balanced diet, and adequate sleep are essential for optimal health and well-being.

ANTI-DOPING CONSIDERATIONS

Peptides have rapidly gained traction in the health and fitness world due to their promising effects on muscle growth, recovery, and overall performance enhancement. However, it's essential to consider the implications of peptide use in professional sports and competitions. Many sports governing bodies, including the World Anti-Doping Agency (WADA), have placed certain peptides on their prohibited substances list. This means that athletes, even if they are using peptides for legitimate medical reasons, can face bans, fines, and other disciplinary actions if these substances are detected in their system during anti-doping tests.

For athletes or anyone participating in regulated sports, it's crucial to be thoroughly informed about the specific peptides or substances that are banned and undergo regular checks. Even peptides available as over-the-counter supplements or commonly used in medical treatments can fall under these restrictions. The challenge lies in the fact that the list of prohibited substances is continuously updated, reflecting ongoing research and emerging knowledge about these compounds. Therefore, staying updated, consulting with sports nutritionists, and always checking any supplement or medication for banned substances are essential steps to ensure compliance with anti-doping regulations.

This chapter has equipped you with foundational knowledge on peptide therapy, including how to get started, what to expect, and potential side effects to be aware of. In the next chapter, we will delve into the specifics of creating your personalized peptide regimen, guiding you to make the most of this promising therapeutic approach.

6

BUILDING YOUR PERSONAL PEPTIDE PROTOCOL

Personalized medicine is increasingly heralded as the future of healthcare, offering treatments and strategies customized to an individual's unique genetic makeup, lifestyle, and medical history. In this evolving landscape, peptides hold significant promise as versatile components of a personalized health strategy. This chapter aims to arm you with the essential information and practical tools needed to create a peptide therapy regimen tailored to your unique health requirements and lifestyle preferences.

By the time you finish reading this chapter, you'll be well-equipped to sift through the plethora of information available, discerning reputable sources from less reliable ones. Furthermore, you'll be prepared to engage in meaningful, well-informed discussions about how peptide therapy could specifically benefit your overall health and well-being. You'll be empowered to take a more active role in managing your healthcare journey.

The focus here isn't just on providing information; it's about enabling you to participate in your healthcare proactively. You'll learn how to navigate the increasingly complex world of personalized medicine, and you'll gain insights into how peptides can be integrated into your health plan for optimal results.

WHY INDIVIDUALITY MATTERS IN PEPTIDE THERAPY

Imagine a healthcare tailored just for you, where treatments fit as perfectly as a pair of bespoke shoes, addressing your unique biological needs rather than following a generic protocol. This is the promise of personalized medicine, and the field of genomics is at its heart. Genomics delves deep, looking at isolated genes linked to certain diseases and exploring the intricate dance of all genes and how they interact with environmental and lifestyle factors. This comprehensive understanding allows healthcare professionals to pinpoint potential health risks and devise strategies tailored to each individual's genetic makeup.

However, we're still navigating the complexities of this approach. As genome sequencing becomes more integrated into healthcare, concerns around data privacy and equitable access emerge. Yet, the transformative potential of personalized medicine can't be understated, promising a future where treatments align seamlessly with individual genetic blueprints.

Peptide therapy, within this context, emphasizes individualization. Each of us has a unique health blueprint shaped by genetics, environment, lifestyle, and past medical experiences. Peptide therapy acknowledges these nuances, avoiding a one-size-fits-all approach. This individualized strategy boosts efficacy and minimizes potential side effects, ensuring safer outcomes.

Beyond treating existing health issues, peptide therapy is proactive. It assesses one's genetic predispositions, formulating strategies to mitigate potential health concerns in the future. The approach empowers patients, fostering a more active role in their health decisions.

As research progresses, our understanding of how individuals respond to peptides continues to deepen. This evolving knowledge ensures that peptide therapy remains attuned to each person's unique needs, pushing us closer to a patient-centric healthcare model that prioritizes safety and effectiveness.

When selecting peptides for therapy, several important factors need to be considered. These elements are essential to help healthcare providers and patients make well-informed choices about which peptides are best suited for personalized treatment:

- **Age:** Different age groups may require distinct peptides to address specific needs. For instance, younger individuals might seek peptides that support youthful skin, muscle growth, or cognitive enhancement. Conversely, older adults might choose peptides targeting age-related concerns such as collagen degradation, joint discomfort, or cognitive decline.
- **Fitness level:** Athletes and those who engage in regular physical activity may opt for peptides like growth hormone-releasing peptides (GHRPs) or BPC-157, which are known to improve athletic performance, muscle recovery, and endurance. Individuals participating in high-intensity physical exercise might also look into peptides that facilitate quicker muscle repair and minimize inflammation.
- **Health goals:** Different peptides can be chosen depending on one's objectives. For example, melanotan could be considered for appetite suppression and fat reduction. Meanwhile,

those dealing with chronic conditions may explore peptides that have shown potential in managing specific issues, such as insulin regulation in diabetes or thymosin alpha-1 (Tα1) for immune support.
- **Current health status:** For those with chronic illnesses, peptides that target the underlying causes or alleviate symptoms might be more appropriate. Individuals recovering from physical injuries could explore peptides like BPC-157 or TB-500, known for their potential to speed up healing and relieve discomfort. People facing mental health issues may consider peptides like noopept, which is thought to have cognitive benefits.
- **Gender:** Gender-specific health concerns may guide the selection of peptides, especially those that address hormonal imbalances.
- **Genetics:** A person's unique genetic composition can affect their susceptibility to particular medical conditions and their response to specific peptides. Genetic screening can, therefore, be instrumental in guiding personalized peptide therapy.
- **Overall health and medications:** The presence of existing health conditions and the use of other medications are essential considerations

in peptide selection to avoid potential adverse interactions.

- **Budget and accessibility:** Finally, the cost of peptides and their availability can also impact an individual's choices, as some peptides might be expensive or hard to access in certain areas.

Choosing the suitable peptides for therapy or supplementation is a nuanced process that requires a personalized approach. Various individual factors, including age, level of physical fitness, specific health goals, and current medical condition, play a critical role in determining the most appropriate peptides. By carefully evaluating these elements, you can make better-informed choices, optimizing the safety and effectiveness of peptide therapy for each individual's unique circumstances.

CREATING A PERSONALIZED PEPTIDE REGIMEN

Think about the following two scenarios:

1. John's Scenario (45 years old, physically active, suffers from chronic knee pain)

At 45 years old, John is at an age where physical wear and tear is becoming more noticeable. Given his active lifestyle, his chronic knee pain is most likely a result of continued physical activity. His main objective is pain relief and promoting healing in his knee joint. Based on these specific needs, certain peptides may be more beneficial for him:

> a. BPC-157: Known for its potential to expedite tissue repair and reduce inflammation, this peptide could be a good fit for alleviating John's chronic knee pain and helping him recover more quickly after intense physical activity.
> b. TB-500: This peptide is another option promoting tissue repair and regeneration. It could help John maintain his physically active lifestyle while minimizing the knee pain troubling him.

Given John's unique health profile and goals, these peptides could be particularly effective in addressing his chronic knee pain, allowing him to continue engaging in the physical activities he enjoys.

2. David's Case (58 years old, corporate professional, concerned about diabetes and weight management)

David is 58 and works in a corporate setting, facing the dual challenges of managing diabetes and weight. His primary health goals are to control his blood sugar levels better and to lose weight. Based on these specific needs, the following peptides may be more suitable for him:

    a. Semaglutide: This peptide has been demonstrated to help with weight loss and blood glucose regulation. It can benefit individuals like David dealing with diabetes and weight issues, providing a dual benefit.

    b. Thymosin Alpha-1: Known for its potential to modulate the immune system, Tα1 could benefit David, especially given his diabetes. It may help support his overall immune health, which is vital for effective diabetes management.

Given David's unique health concerns and objectives, these peptides are well-aligned with his focus on diabetes control and weight management, two issues that are often closely related.

The therapeutic peptide choices for people like John and David exemplify how personalized medicine tailors treatment to specific health needs and goals. Their age, lifestyle, and health conditions guide the selection of peptides, ensuring they receive targeted and effective solutions tailored to their circumstances. To choose the right peptide therapy for yourself, follow the following steps:

1. Self-evaluation

Conducting regular health checkups and diagnostic tests is a crucial first step for anyone considering peptide therapy. These evaluations serve multiple purposes. Firstly, they offer a foundational assessment of your overall health, acting as a baseline against which future changes can be measured. Secondly, they can identify current medical conditions, potential risk factors, or specific health issues requiring intervention. With a clear understanding of your health status, you can make well-informed decisions about whether peptides are an appropriate addition to your healthcare regimen.

People vary in their genetic makeup, lifestyle, and health background. Self-evaluation takes these individual factors into account. It helps you recognize how genetics, diet, exercise, and stress affect your health, allowing you to customize your peptide regimen accordingly. It enables you to identify any contraindications or potential interactions between peptides and your existing medications or health conditions. This knowledge is critical to ensure your peptide therapy is safe and effective. To perform self-evaluation:

- **Schedule regular checkups:** Consistent medical checkups provide a baseline for understanding your health status. These visits typically include physical examinations, blood tests, or other diagnostic assessments. Three months' worth of checkup data can offer valuable insights.

- **Examine medical history:** Review your medical history, including any chronic illnesses, past surgeries, or other major health events. This information helps healthcare providers assess how peptides could interact with your existing medical conditions and optimize your treatment plan.
- **Evaluate lifestyle factors:** Assess lifestyle elements such as diet, sleep quality, exercise habits, and stress levels. These variables can significantly influence your overall health and may affect which peptides are most beneficial for you. Providing this information allows for a more tailored approach to peptide therapy.

Maintaining a well-organized record of your health data is a vital part of self-evaluation. Make sure to document your test results, observations from regular checkups, and any noticeable changes in your health. This comprehensive record can be a valuable tool for you, aiding in tracking progress and adjusting your peptide therapy plan as needed.

Regular medical checkups and diagnostic tests are foundational to developing a personalized peptide regimen. By allowing you to make well-informed decisions, these evaluations help to clarify your health goals. Consequently, you can establish a peptide therapy plan that is effective, safe, and tailored to meet your unique health needs and objectives.

2. Identify and clarify your health goals.

Setting well-defined health goals is pivotal to optimizing your peptide therapy. Clear objectives ensure that the chosen peptides match your desired outcomes: managing a specific illness, enhancing athletic performance, or making aesthetic changes. Different peptides offer distinct benefits, so clarity in your objectives ensures your therapy is tailored to your unique needs.

By having precise goals, you can monitor your progress more efficiently. Defined objectives provide benchmarks to evaluate the therapy's effectiveness; adjustments can be made for better results if necessary. A clear vision of your intended outcomes also enhances commitment and motivation, leading to better adherence to your regimen and any supportive lifestyle changes.

Additionally, having clear goals bolsters communication with healthcare professionals. When they understand your objectives, they can offer better guidance, monitor your progress, and make informed recommendations. Reflecting on the specific health areas you aim to enhance or maintain is beneficial. Whether your aspirations are weight loss, muscle building, cognitive improvements, pain reduction, or skin rejuvenation, documenting these goals reinforces commitment. If multiple objectives are identified, it's crucial to prioritize them. For example, focusing on weight loss before muscle toning can help allocate resources and attention more effectively.

3. Consult with specialists

Consulting with specialists, be it endocrinologists, dermatologists, or sports medicine professionals, can provide targeted insights tailored to your peptide regimen. These experts can offer a thorough risk assessment, considering any existing health conditions, medications, or genetic factors that might affect peptide therapy's safety and efficacy. Their expertise ensures you make well-informed choices regarding suitable peptides, their dosages, and frequency.

With peptide therapy, continuous monitoring is often required, and specialists can adeptly track your progress, perform follow-up assessments, and make treatment plan adjustments. For those with multiple health conditions or taking several medications, these experts can ensure that peptide treatments harmoniously integrate with other healthcare interventions, minimizing potential conflicts or interactions.

Some specialists might suggest advanced peptides unavailable over the counter, guiding you through the benefits, potential side effects, and anticipated outcomes of such treatments. Engaging openly, asking questions, and adhering to their recommendations are crucial to making the most of these consultations, as they pave the way for a successful, personalized peptide regimen.

ADVANCED STRATEGIES FOR OPTIMIZED PEPTIDE THERAPY

Peptide therapy should never replace living a healthy lifestyle; it should supplement it. Use it alongside the following diet and lifestyle recommendations to get the best out of it:

- **Intermittent fasting**

If you have tried to lose weight before, you know that it is not always straightforward. Intermittent fasting has become a popular weight loss tool. The concept has captured the attention of many, and for good reason. There are various approaches to intermittent fasting, each boasting the potential to yield powerful and effective results for weight management and promoting healthy aging and disease prevention. While many people have experienced significant weight loss through intermittent fasting alone, there are cases where a complementary approach is necessary. Incorporating carefully chosen peptide therapies into an intermittent fasting regimen can be a game-changer, leading to remarkable weight loss outcomes.

The advantages of fasting go beyond the simple notion that reducing calorie intake results in weight loss. Fasting does lead to calorie reduction, but it also triggers profound physiological changes within the body. These changes facilitate the burning of stored fat and activate genes responsible for regulating metabolism, enhancing insulin sensitivity, improving metabolic efficiency, and promoting cellular rejuvenation. Within our cells, the regular metabolic process generates waste products that must be either disposed of or recycled by our bodies. This cellular cleansing process is closely associated with overall health and longevity. Fasting is pivotal in activating the genes that oversee these crucial cellular mechanisms.

Our distant ancestors lived in an era where food was often scarce. Over time, our genetic makeup adapted to cope with these lean periods. This adaptation mirrors the behavior of hibernating mammals, which accumulate fat during the summer months to harness stored energy and rejuvenate during the winter. Intermittent fasting taps into this. It isn't just a dietary trend; it reaches into our innate genetic resilience, offering a pathway to weight control, improved health, and a rejuvenated sense of well-being. It involves abstaining from consuming calories for a significant duration, typically spanning 24 hours or more. There are several intermittent fasting programs:

- **5-2 fasting program:** This regimen involves five days of normal eating and two non-consecutive 36-hour fasting periods weekly. For example, you could stop eating after dinner on Sunday and resume with breakfast on Tuesday. This pattern can also be applied to other days, like fasting from Thursday dinner to Saturday breakfast.
- **Extended consecutive fasting:** This more intensive option involves fasting for three to five days in a row, typically undertaken once or twice a month.
- **Fasting-mimicking diet:** This five-day plan allows for consuming specific, limited-calorie foods. Even though you consume some calories, the goal is to induce metabolic changes similar

to those achieved by fasting. The frequency of this program varies depending on individual objectives.

- **Time-restricted eating:** This approach confines daily caloric intake to a specific time window, such as an 8-hour period. The popular 16/8 schedule involves eating between, for instance, noon and 8 p.m. and fasting for the remaining 16 hours. The time window can be adjusted according to personal preference and lifestyle, and some people initiate the program with a more lenient 14/10 schedule before progressing to a narrower eating window.

The popularity of these fasting programs can be attributed to their flexibility, potential health benefits, and adaptability to various lifestyles. You can tailor them to individual preferences and goals and embark on a path to better health and well-being. It's important to note that these eating schedules place less emphasis on specific food choices and focus primarily on meal timings. When it's time to eat, you can relish your meals without constant worry about dietary restrictions, but you still need to maintain proper nutrition. Prioritize whole foods, avoid added sugars and processed foods, and limit refined carbohydrates. Be mindful not to overindulge during the designated feeding period.

Intermittent fasting can be challenging; not everyone can commit to extended fasting. For many, adhering to a time-restricted eating schedule, even a 14/10 program, is more feasible. Some individuals may not achieve their weight loss goals through intermittent fasting alone. Besides, the aim should be to lose fat while preserving muscle, which can be complicated, especially with caloric restriction.

This is where peptides come into play. Several peptides can serve as valuable additions to an intermittent fasting program, enhancing its effectiveness and supporting overall health:

- **Growth Hormone-Stimulating Peptides**

Peptides such as ipamorelin, CJC-1295, sermorelin, and tesamorelin are often used to stimulate the production of growth hormone (GH) by the pituitary gland. The production of GH naturally declines with age, leading to muscle loss, increased fat accumulation, reduced bone density, and other age-related changes. While direct supplementation with human growth hormone (HGH) is generally not recommended due to safety concerns, these peptides serve as a safer alternative by mimicking the naturally occurring peptides that trigger GH production in the pituitary gland.

Incorporating these GH-stimulating peptides into an intermittent fasting regimen can be beneficial for achieving health objectives, particularly in terms of preserving muscle mass and promoting general wellbeing. Used on their own, these peptides have been shown to improve body composition by reducing fat and increasing muscle mass, although their effect on overall weight loss may be limited. When combined with intermittent fasting, the peptides may enhance the regimen's effectiveness, aiding muscle preservation and strength improvement.

- **Liraglutide**

Liraglutide, marketed under the trade name Victoza, has served as a peptide medication for diabetes management since 2010. Alongside several similar diabetes drugs, such as Trulicity, Ozempic, Bydureon, and Byetta, these peptide-based products function as "incretin analogs." They mimic intestinal hormones responsible for regulating carbohydrate metabolism, appetite, metabolism, and intestinal function. Initially designed for Type 2 diabetes treatment, these medications also demonstrated remarkable weight loss effects in treated patients.

Liraglutide is an effective weight loss agent for non-diabetic people and received approval as Saxenda in 2014. Saxenda involves daily injections and can benefit intermittent fasting or other weight loss dietary regimens. It doesn't induce hypoglycemia but diminishes appetite and encourages early satiety, reducing the tendency to overeat. In clinical studies involving non-diabetic subjects, liraglutide users achieved an average weight loss of 8% of their total body weight.

- **Mediterranean diet**

You've likely come across the term Mediterranean diet at some point. This dietary approach is frequently touted for its potential to reduce the risk of heart disease, depression, and dementia.

The Mediterranean region encompasses various countries with slightly different dietary traditions, resulting in several versions of the Mediterranean diet. In 1993, a collaborative effort between the Harvard School of Public Health and other organizations led to the introduction of the Mediterranean Diet Pyramid. This pyramid served as a guide to familiarize people with the prevalent foods of the Mediterranean region. More than a strict diet plan, it emphasized an eating pattern based on the dietary habits of mid-20th-century Crete, Greece, and southern Italy. During that period, these regions displayed low rates of chronic diseases and above-average life expectancy despite limited access to healthcare. It was believed that their diet contributed to their remarkable health. The pyramid highlighted the importance of daily physical activity and the social benefits of communal dining.

The Mediterranean diet primarily revolves around plant-based foods, including daily consumption of fruits, vegetables, beans, nuts, whole grains, olive oil, spices, and red wine. You consume animal proteins in smaller quantities, preferring fish and seafood as the primary sources. While the pyramid provides a visual guide indicating the proportions of foods to eat (e.g., more fruits and vegetables, less dairy), it doesn't specify precise portions or specific amounts. Portion decisions are left to the individual, considering factors like physical activity level and body size.

The Mediterranean diet stands out because it encourages daily physical activity and emphasizes the social aspects of communal meals, promoting shared dining experiences. It encourages the consumption of red wine in moderation. It offers a flexible and balanced approach to eating, emphasizing whole, unprocessed foods and the healthful traditions of the Mediterranean region.

Extensive research consistently supports the effectiveness of the diet in lowering the risk of cardiovascular diseases and overall mortality. For instance, a study involving nearly 26,000 women revealed that those adhering to this dietary pattern experienced a 25% lower risk of developing cardiovascular disease over 12 years. The study looked into various underlying mechanisms contributing to this risk reduction and identified changes in inflammation, body mass index, and blood sugar levels as significant factors. Similar benefits were observed in a meta-analysis of 16 prospective cohort studies encompassing over 22,000 women monitored for approximately 12.5 years. Those with the highest adherence to the diet exhibited a 24% lower risk of cardiovascular disease and a 23% lower risk of premature death compared to those with the lowest adherence.

A wealth of research supports the Mediterranean diet as a healthy eating pattern for preventing cardiovascular diseases, extending lifespan, and promoting healthy aging. When combined with caloric restriction, this diet can help in maintaining a healthy weight.

The diet can complement and enhance the effects of peptide therapy. Adopting this dietary pattern alongside specific peptides, you may experience synergistic benefits that support overall well-being. The

Mediterranean diet is known for its anti-inflammatory qualities, which can be amplified when combined with specific peptides. For instance, peptides like BPC-157 and TB-500 have demonstrated anti-inflammatory properties. When taken alongside a diet rich in fruits, vegetables, and olive oil, these peptides may provide enhanced relief from inflammatory conditions and promote quicker healing.

Peptides such as CJC-1295 and ipamorelin can contribute to heart health by promoting better cholesterol profiles and reducing the risk of heart disease. Coupled with the Mediterranean diet's emphasis on healthy fats, particularly olive oil and omega-3-rich fish like salmon, people can further enhance their cardiovascular well-being.

The Mediterranean diet is also recognized for its role in healthy weight management. Combining it with weight management peptides like AOD-9604 or tesamorelin may yield more effective weight loss results. These peptides can help you shed excess fat while preserving muscle mass, aligning with the Mediterranean diet's focus on lean proteins and whole foods. The diet is rich in antioxidant-rich foods such as fruits, vegetables, and nuts. Combined with peptides like glutathione or thymosin beta-4, which have antioxidant properties, you can enhance your body's ability to combat oxidative stress and reduce cellular damage.

It's important to note that combining peptide therapy with any diet should be done under the guidance of a specialist. Individual needs and health conditions vary, so a personalized approach is essential to ensure safety and maximize the benefits of diet and peptide therapy. Always consult an expert before starting any new dietary or therapeutic regimen.

- **Moderate exercise**

Moderate exercise, such as walking, combined with strength training, can significantly boost the effectiveness of peptide therapy. The synergy between physical activity and peptides can enhance overall health benefits. Peptides like PT-141 and CJC-1295 improve blood flow and vascular health. Moderate exercise, particularly aerobic movements like walking, naturally promotes better circulation. Combined, they can collectively support cardiovascular health by lowering the risk of clot formation and enhancing the delivery of peptides to target tissues.

Exercise helps weight management by burning calories and increasing metabolic rate. Combined with peptides like AOD-9604 or tesamorelin, which promote fat loss while preserving muscle mass, you can experience more effective and sustainable weight loss results. Peptides like BPC-157 and TB-500 are known for their tissue healing and recovery properties. Moderate exercise can help maintain joint mobility and flexibility. Combined with these peptides, you may experience faster recovery from exercise-induced strain or injuries.

Exercise has well-documented mood-enhancing effects. Peptides like selank and noopept can further support cognitive function and emotional well-being. Together, they can contribute to an improved overall sense of well-being. Remember that fitness levels and health goals vary, so any exercise regimen should be tailored to individual needs.

- **Strength training**

When incorporated into a fitness routine for both women and men, strength training can yield numerous benefits that complement and enhance the effects of peptide therapy. Strength training, including resistance exercises and weightlifting, promotes muscle growth and maintenance. Peptides like ipamorelin, sermorelin, and GHRP-6 stimulate growth hormone release, supporting muscle development. When used alongside strength training, these peptides can amplify muscle growth, increase strength, and improve metabolism.

Resistance training stimulates bone growth and increases bone density. It places stress on the bones, prompting the body to fortify them. Peptides like ostarine (MK-2866) have shown potential for promoting bone health. Strength training can further enhance your bone density, reducing the risk of osteoporosis and fractures (especially for women approaching menopause and older adults).

Since strength training significantly influences metabolic function, it burns calories during exercise and enhances post-exercise calorie expenditure. Adding peptides like AOD-9604 or tesamorelin can further support the metabolic process by promoting fat loss while preserving muscle mass. This combination can be particularly beneficial for people aiming to improve their metabolic health and lose weight.

Strength training can enhance joint stability and function when performed correctly and under guidance. This is particularly beneficial for those with joint-related issues. Peptides like collagen peptides can further support joint health. Together, they can promote joint flexibility and reduce the risk of injuries. It's important to note that strength training programs should be fitted to individual fitness levels and goals.

Before starting any exercise regimen or peptide therapy, consult with a healthcare provider or fitness expert to ensure safety and optimize the benefits of both approaches.

- **Daily meditation and gratitude journaling**

Daily meditation and gratitude journaling can complement and amplify the effects of peptide therapy by promoting mental and emotional well-being, reducing stress, and enhancing overall health. Integrating mindfulness practices with peptide therapy offers a comprehensive approach to overall well-being. Meditation is an excellent tool for stress reduction. It helps you manage your response to stressors and fosters a sense of calm. High-stress levels can negatively impact overall health and interfere with the effectiveness of peptide therapy. You may experience enhanced receptiveness to peptide treatments by reducing stress through meditation.

Daily meditation can also lead to improved mood and emotional well-being. Peptides like selank and noopept are known for their mood-enhancing properties. Combined with meditation, these peptides can contribute to a more positive outlook, reduced anxiety, and a greater sense of contentment.

Gratitude journaling is about writing down things you are thankful for daily. This practice fosters a positive mindset and can improve overall well-being. Peptides like selank and thymosin alpha-1 are known for their immune-boosting properties. When used alongside gratitude journaling, these peptides can support a healthier immune system, reducing the risk of illness.

Meditation and gratitude journaling promote a solid mind-body connection, making you more attuned to your physical sensations and health needs. This heightened awareness can enhance the perception of how peptide therapy affects the body, making it easier to fine-tune treatment plans for optimal results. Combining meditation, gratitude journaling, and peptide therapy creates a holistic approach to wellness. It addresses physical, mental, and emotional health, contributing to a more balanced and fulfilling life.

IDENTIFYING RELIABLE SOURCES OF INFORMATION

It's crucial to consult reliable information sources to make informed decisions about peptide therapy. Here are some guidelines and resources to help you:

- **Verify author credentials:** Seek articles written by qualified individuals, such as medical professionals or scientists, to ensure credibility.

- **Use reputable health websites:** Well-known health and medical platforms like the Mayo Clinic, WebMD, Harvard Health Blog, and MedlinePlus offer credible information.
- **Consult scientific journals:** Academic journals, such as the Journal of Peptide Research, are excellent sources for up-to-date, peer-reviewed peptide research.
- **Talk to healthcare professionals:** Always consult healthcare providers for personalized, expert advice before deciding on peptide therapy.
- **Review peer-reviewed studies:** Look for research articles in peer-reviewed journals. PubMed is a valuable resource for this type of information.
- **Webinars and conferences:** Organizations like the American Peptide Society often host events featuring discussions on the latest peptide research.
- **Clinical trials:** Websites like ClinicalTrials.gov offer information on ongoing clinical trials, which can provide insights into the effectiveness of different peptide therapies.

- **Refer to government health agencies:** Governmental bodies like the National Institutes of Health (NIH) and the Food and Drug Administration (FDA) often provide verified information on approved peptide therapies and associated safety guidelines.

As a rule of thumb, be wary of sources promising miraculous, instant results. Genuine health improvements usually require time and often a combined approach. Always check references when consulting online articles. If you consult pharmaceutical company websites, cross-reference their information with independent sources to ensure objectivity. Avoid relying solely on one source for information. Remember that medicine continually evolves, and new research may emerge over time.

YOUR PEPTIDE THERAPY CHECKLIST

Use this checklist to organize and track progress throughout your peptide therapy journey. It will help you document your health goals, chosen peptides, and lifestyle changes, monitor your progress and side effects to stay on track, and work collaboratively with your healthcare provider to optimize your therapy plan.

Health goals

- Define your specific health goals and objectives for peptide therapy.
- Identify the primary areas of improvement you seek (e.g., weight management, muscle gain, joint health, anti-aging, etc.).

Chosen peptides

- Consult a specialist to determine the most suitable peptides for your goals.
- List the names of the selected peptides and their recommended dosages.
- Keep track of the start date for each peptide.

Lifestyle changes

- Consider necessary lifestyle modifications that complement your peptide therapy plan (e.g., diet, exercise, sleep, stress management).
- Incorporate these changes into your daily routine.

Monitoring progress

- Regularly assess your progress toward your health goals.
- Use objective measurements (e.g., weight, body composition, blood pressure) to track improvements.
- Document any positive changes or improvements in your health.

Side effects and reactions

- Be aware of potential side effects associated with your chosen peptides.
- Record any side effects or adverse reactions experienced during the therapy.
- Consult with your healthcare provider if you encounter side effects.

Compliance

- Maintain a consistent schedule for taking your peptides as prescribed.
- Ensure you follow the recommended dosages and administration instructions.
- Keep a record of missed doses, if any.

Consultations

- Schedule regular follow-up appointments with your healthcare provider to discuss progress and make any necessary adjustments to your peptide therapy plan.
- Share your checklist and updates with your healthcare provider during appointments.

Notes and observations

- Use this section to write additional observations, thoughts, or questions about your peptide therapy journey.

This chapter gave you the knowledge and tools to embark on a personalized peptide therapy journey tailored to your unique health needs and goals. You have seen that personalized medicine is the future, and peptides offer an exciting avenue to customize your healthcare. You understand that one size doesn't fit all, and your peptide therapy should align with your requirements.

You now know to consider your age, fitness level, health goals, current health status, and other factors significantly influencing your peptide choices. Choose peptides that best suit your specific circumstances.

Peptides are versatile, but not all are suitable for every purpose. Now that you understand how to implement peptide therapy, it's time to put your knowledge into action. Create your personalized peptide protocol, considering your health goals and individual factors. Consult healthcare professionals to refine your plan and ensure its safety and effectiveness.

The following chapter delves into the relationship between biohacking and peptides. It will provide insights into how these two areas intersect, offering you tools to manage and optimize your health proactively.

7

PEPTIDES MEET BIOHACKING: A SYNERGY FOR SUPERIOR HEALTH

Biohacking offers a systematic approach to enhancing one's well-being and performance. Through this scientific method of self-improvement, individuals can optimize their body's functions, pushing the boundaries of what's traditionally considered the limits of human potential.

In this chapter, we will explore the relationship between peptides and biohacking. You'll gain a clear understanding of the principles of biohacking and see how biohackers utilize peptides to enhance health and well-being. With examples from real-life applications and insights from leading experts, this chapter aims to provide a comprehensive view of peptides' potential within the biohacking realm. This in-depth exploration will equip you with the knowledge to make informed decisions about integrating peptides into your biohacking journey.

WHAT EXACTLY IS BIOHACKING?

Biohacking involves making science-informed adjustments to one's lifestyle to enhance health. These adjustments can encompass dietary changes, daily routines, or supplements tailored to an individual's needs. Since everyone's body and circumstances are unique, a strategy effective for one person might not be for another. It's essential to prioritize biohacking methods supported by scientific evidence and approach unverified techniques cautiously.

Biohacking encompasses a range of practices individuals adopt to optimize their health and well-being. Motivations for biohacking vary: some seek better control over their health, others are curious about new wellness techniques, while some aim for longevity or self-improvement. Biohacking is a proactive, self-directed approach to improving one's biology. It can involve dietary changes, lifestyle modifications, cognitive enhancements, or accelerated weight loss strategies. While some biohacks are safe and can be tried independently, others might carry risks or yield unpredictable outcomes, underscoring the importance of research and caution.

Biohacking often incorporates nootropics, substances intended to improve cognitive functions. These can be found in pills, supplements, and even certain foods and drinks, with the aim of enhancing mental performance. Typical examples are caffeine and creatine. Some nootropics are prescription medications used to treat conditions like Alzheimer's disease or attention deficit hyperactivity disorder (ADHD). Examples of these prescription nootropics are methylphenidate (known as Ritalin), Adderall, and memantine (marketed as Axura).

Using prescription medications should always be in accordance with a doctor's guidance. Stimulants, when misused, can lead to a range of side effects. Studies indicate that non-medical use of prescription stimulants can increase the risk of anxiety, heightened use of other substances, post-traumatic stress disorder, engaging in unsafe sexual practices, and worsening academic performance.

Wearable technology plays a significant role in the realm of biohacking. Devices such as smartwatches, fitness trackers, and head-mounted displays have become increasingly popular, allowing individuals to continuously monitor a wide range of health metrics. By analyzing this data, users can make more informed health and fitness decisions, set and achieve specific milestones, or even monitor intricate patterns like reproductive health cycles.

Beyond these wearable gadgets, the technological frontier is pushing the boundaries with embedded implants. These are surgically positioned devices within the body with various functionalities. Leading technology firms are pioneering this domain, exploring implants that can store personal data, act as electronic keys for secure access, and provide deep insights into an individual's physiological metrics. Among the innovations in this category are magnetic units that react to external stimuli, memory chips for data storage, GPS modules for location tracking, and electronic tattoos that merge tech with aesthetics. All these advancements highlight the expanding and evolving intersection of biohacking and wearable technology.

Generally, there are three types of biohacking:

- **DIY biology**

DIY biology, often called garage biology, involves individuals or groups conducting scientific research outside of professional settings. In this approach, knowledgeable individuals share scientific techniques and information with enthusiasts who might not have formal training. Doing so democratizes access to biological research, allowing for experimentation outside established institutions. Some see DIY biology as a way to challenge traditional scientific gatekeeping, emphasizing inclusivity and community-driven research. The activities under DIY biology can vary widely, spanning areas like microbiology, nutrition, and even advanced fields like synthetic biology.

- **Nutrigenomics**

Nutrigenomics studies the interaction between food and an individual's genetic makeup. This field investigates how different foods may affect a person based on their genes. By exploring the connections between diet, genetics, and health outcomes, researchers hope to devise personalized disease prevention and treatment strategies. In practice, individuals provide a DNA sample for analysis. Once their genetic profile is determined, they receive tailored nutritional advice. This advice often includes recommendations on foods to consume or avoid, especially if certain foods could exacerbate conditions the individual is genetically inclined toward.

- **Grinder movement**

The grinder movement is a subset of the biohacking community focused on human augmentation by implanting devices and technology under the skin. Grinders seek to enhance human capabilities by integrating technology directly into the body. Their practices often involve embedding devices to push the limits of human biology and technological fusion.

The origins of biohacking are not easily traceable. Pinpointing the first instance of someone modifying organisms like E. coli outside a professional lab setting remains uncertain. Some suggest that Rob Carlson might have been one of the early pioneers in the mid to late 2000s. The term "biohacking" has seen various interpretations, and its exact definition has evolved over time. Historically, a notable event was the 2004 arrest of artist Steve Kurtz, who worked with bacteria and yeast. Kurtz's endeavors bear similarities to modern-day biohacking concepts.

Biohacking encompasses a wide range of practices and interpretations. Many viewed the term with skepticism in its early stages, and it wasn't widely adopted. Some, including Hank Greely, even see "biohacker" as a derogatory label. Initially, "DIY Biologist" was more commonly accepted as it seemed less controversial.

However, this term has been criticized by some who believe it suggests that a scientist working outside a traditional lab setting is somehow less competent or not a "real" biologist.

Biohackers have positioned themselves at the forefront of experimental fields such as medicine, surgery, and genetics. Human experimentation requires rigorous documentation, ethical review processes, and institutional approvals in traditional academic settings. However, biohackers have identified a unique space for themselves; by focusing on self-experimentation, they bypass the formalities and approvals typically required by organized institutions. This approach allows them to explore without the constraints of traditional research environments.

HOW BIOHACKERS ARE USING PEPTIDES TO GET HEALTHIER

Biohackers increasingly turn to peptides as a potent tool to optimize their health and well-being. They are leaning into the fact that peptides play crucial roles in various physiological processes within the body. Here are some popular peptides and their uses in the biohacking community:

- **Semaglutide for weight management**

Semaglutide, initially developed by Novo Nordisk in 2012, gained FDA approval in 2021 for weight management. It belongs to the class of medications called GLP-1 agonists. Semaglutide stimulates the pancreas to increase insulin production, which slows down stomach emptying, leading to feelings of fullness. This appetite suppression supports gradual and sustainable weight loss over several months without requiring invasive surgeries or stimulants.

- **CJC-1295/ipamorelin for various benefits**

CJC-1295/ipamorelin is a peptide combination used for several purposes, including increasing lean muscle mass, reducing inflammation, boosting the immune system, improving sleep quality, enhancing strength, and increasing collagen and elastin production. It may also increase energy levels, sex drive, and cardiac health and accelerate healing after injuries.

This peptide combination works synergistically to stimulate higher levels of growth hormone production. It is often used in daily subcutaneous injections for five days straight, followed by two days off, over a 12-week treatment program. You begin to notice positive changes, such as improved sleep and subtle body composition changes, within the first few weeks, with the best results typically occurring around month three.

- **Epitalon for anti-aging**

Epitalon is being used for its potential anti-aging effects. It can lengthen telomeres, promoting cell replication and rejuvenation. This peptide is also associated with improved sleep quality, reduced inflammation, boosted immune function, slowed aging, potential cancer protection, increased cell rejuvenation, and lowered risk for dementia. It's a potent antioxidant, regulates insulin and cholesterol levels, and accelerates wound healing.

- **GHK-Cu for tissue repair and anti-aging**

GHK-Cu is a copper peptide recognized for its diverse functions in the human body. It aids in wound healing, recruits immune cells, exhibits antioxidant and anti-inflammatory effects, encourages the production of collagen and glycosaminoglycans in skin cells, and fosters the growth of blood vessels. Playing a significant role in tissue repair and protection, GHK-Cu levels naturally diminish as we age. This decline can be associated with heightened inflammation, greater tissue vulnerability, and other age-linked challenges.

These peptides offer biohackers a range of tools to optimize their health, from achieving weight management goals to enhancing muscle mass, improving sleep quality, and potentially slowing down the aging process.

THE BIOHACKING CULTURE

In an interview with Dr. Heather Sandison, Dave Asprey, a well-known figure in the field of biohacking, discussed his interest in peptides and SARMs (Selective Androgen Receptor Modulators) as part of his journey toward achieving optimal health and longevity. According to him, peptides are signaling molecules in the body that play a crucial role in various physiological processes. Asprey simplifies them by comparing them to words made up of letters from the alphabet. In this analogy, amino acids are the alphabet of proteins, peptides are individual words, larger peptides form sentences, and complete proteins become paragraphs or pages. Asprey's fascination with peptides lies in their potential to influence different aspects of health and well-being.

One notable example of a peptide he used is collagen, which he introduced to the market through his Bulletproof brand. Collagen is a protein made up of di and tripeptides, which Asprey highlights as being particularly effective for various health benefits, including anti-aging and performance enhancement. He contends that these peptides play roles in functions like lengthening telomeres and speeding up healing, suggesting they might offer benefits beyond some conventional pharmaceuticals.

Asprey recognizes the high cost of peptides, ranging from $50 to $300. However, he stresses their potential advantages in anti-aging and overall health enhancement. Dr. Heather Sandison, during the interview, also advocates for peptides, particularly thymosin beta-4 and thymosin alpha-1. She considers them transformative for her patients due to their effectiveness, affordability, and low-risk profile.

You can see examples of people embracing biohacking even in the documentary film Citizen Bio, produced by Showtime. The film highlights many biohackers and their pursuit of medical vigilantism. Released in 2020, this documentary closely examines people pushing the boundaries of traditional medicine and science, often using unconventional methods and substances, including peptides, to achieve their health and wellness goals.

It explores the stories of several biohackers who are determined to take control of their health and extend their lifespans through self-experimentation. These biohackers are driven by a desire to transcend the limitations of conventional medicine and scientific research, and they often resort to innovative and sometimes risky approaches to achieve their objectives.

Among their biohacking tools of choice are peptides. The documentary shows biohackers experimenting with peptides, among other substances, in their quest for self-improvement. Biohackers believe that peptides hold the potential to unlock hidden abilities, reverse aging, and enhance physical and mental performance. While some of their experiments are unorthodox and raise ethical questions, they are driven by a desire to take control of their biology and challenge the healthcare status quo.

Citizen Bio is a thought-provoking exploration of the biohacking subculture and the ethical dilemmas associated with self-experimentation. It raises questions about the boundaries of scientific research, the regulation of emerging technologies, and the potential risks and rewards of people taking their health into their own hands.

While some may view biohackers as medical vigilantes, they are undeniably at the forefront of a movement that challenges traditional notions of healthcare, offering a glimpse into a future where people have greater control over their well-being. Citizen Bio provides a captivating and sometimes cautionary look into this wild west of biohacking, where the boundaries of medical science are continually tested and redefined.

One wonders what the future holds. Could peptides become a staple in our everyday health routines? We will delve into the exciting possibilities in the next chapter.

8

PEPTIDES: USHERING IN A NEW ERA OF WELLNESS

Imagine a future where the secrets to optimal health, longevity, and vitality lie within our very own biological makeup, specifically in peptides. As we stand on the cusp of a wellness revolution, peptides may be the game-changers we've been waiting for.

Medical science and technology continue to progress rapidly, with peptides emerging as a significant potential contributor to advancements in wellness. This chapter delves into the function of peptides in individual health paths, examines current trends, and presents expert predictions about the future of healthcare. The goal is to provide you with a comprehensive understanding of peptide therapy's potential to change our perspectives on health and well-being.

THE LATEST IN PEPTIDE SCIENCE

Peptides have emerged as a significant player in the pharmaceutical landscape. Once primarily associated with drug development, peptides have expanded their influence into diverse fields such as skin care, fitness, and cognitive enhancement. The rapid growth in their usage across these sectors is indicative of the transformative potential that peptides hold for improving various aspects of our lives:

- **Skincare**

Peptides have become a prominent player in the skincare industry and for a good reason. These short chains of amino acids can address a wide range of skin issues. For instance, collagen-boosting peptides can help reduce the appearance of wrinkles and fine lines. Others, like antimicrobial peptides, are used in acne treatments. Peptide-based skincare products are gaining popularity due to their ability to promote skin repair, hydration, and rejuvenation. Ongoing research continues to explore new peptide formulations for enhanced skincare, making them a vital component of the beauty and wellness market.

- **Fitness**

In fitness, peptides are making waves as potential performance enhancers and tools for muscle recovery. GHRPs stimulate growth hormone release and help in muscle growth and fat loss. Specific peptides are being investigated for their potential to enhance endurance and promote faster recovery post-exercise. The fitness industry is increasingly exploring the benefits of peptide therapies to optimize athletic performance safely and effectively.

- **Cognitive enhancement**

Cognitive enhancement peptides are also gaining attention. Some peptides, like nootropics, are being studied for their potential to boost cognitive function, memory, and focus. These compounds enhance mental clarity and performance without the side effects associated with traditional stimulants. Ongoing research is focused on identifying new peptides and optimizing their formulations to support cognitive health and enhancement.

As research into peptides advances, it's becoming increasingly clear that their applications extend beyond these three areas. Peptides are also being explored for their potential in immune system modulation, weight management, and even personalized medicine. The growth in their usage across diverse fields demonstrates their versatility and underscores their immense potential in shaping the future of health and wellness.

In a testament to their remedial potential, the top 200 drug sales in 2019 included no less than ten non-insulin peptide drugs. Among these, the most noteworthy were GLP-1 analogs designed to treat Type 2 Diabetes Mellitus (T2DM). Trulicity (dulaglutide) secured the 19th position in the sales rankings, with an astonishing $4.39 billion in retail sales. Victoza (liraglutide) followed closely behind, ranking 32nd with $3.29 billion in sales, while Rybelsus (semaglutide) clinched the 83rd spot, boasting $1.68 billion in sales. These figures underscore the significant impact of peptide-based therapeutics in addressing critical health conditions.

The trajectory of peptide research and development continues upward. Peptide drugs have gained increased interest and recognition in recent years, leading to a wave of innovation in medicine. Over 60 peptide drugs have successfully reached the market, benefiting countless patients worldwide. Even more promising is the fact that several hundred novel therapeutic peptides are currently in various stages of preclinical and clinical development. This robust pipeline of potential treatments signifies that the peptide revolution in healthcare is far from reaching its zenith, promising a future where peptides play an even more prominent role in improving human health and well-being.

It's important to note that while the prospects are exciting, peptide-based interventions are still relatively new, and further research is needed to fully understand their long-term effects, safety profiles, and optimal usage. As the wellness revolution driven by peptides continues to evolve, ongoing scientific inquiry will be essential to ensure that these promising developments are harnessed for the betterment of human health.

THE EXCITING FUTURE USES OF PEPTIDES

Therapeutic peptides are emerging as potent tools in managing chronic illnesses and potentially curing diseases, an advancement that marks a significant stride forward in modern medicine. The versatile nature of peptides allows them to be tailored for precise molecular interactions, making them an attractive option for addressing various medical conditions such as:

- **Diabetes mellitus**

Peptides have revolutionized the treatment of diabetes mellitus, particularly Type 2 Diabetes (T2DM). GLP-1 analogs, such as dulaglutide, liraglutide, and semaglutide, are among the most successful peptide-based therapies. These peptides mimic the action of the hormone GLP-1, stimulating insulin release, suppressing glucagon secretion, and promoting glucose control. With their impressive sales figures and efficacy, they have transformed the management of T2DM and hold the promise of better glycemic control for millions of patients.

- **Cardiovascular disease**

Peptides are becoming valuable tools in treating cardiovascular diseases. B-type natriuretic peptide (BNP) analogs, such as nesiritide, have effectively alleviated acute heart failure symptoms by reducing blood pressure, minimizing fluid retention, and enhancing cardiac function. Researchers are investigating new peptide-based treatments targeting various cardiovascular issues, including hypertension and atherosclerosis.

- **Gastrointestinal diseases**

Therapeutic peptides are also finding applications in gastrointestinal diseases. For example, octreotide and lanreotide are somatostatin analogs used to manage conditions like acromegaly and neuroendocrine tumors. They work by inhibiting the release of hormones that contribute to disease progression. Peptide-based therapies continue evolving, offering new treatment options for gastrointestinal disorders, including inflammatory bowel disease.

- **Cancer**

Peptides are gaining attention for their potential in cancer treatments. Peptide-based medications, like GnRH analogs, are applied to treat hormone-dependent cancers, including prostate cancer. Immunotherapy methods using checkpoint inhibitors, such as PD-1 and CTLA-4-blocking peptides, are also emerging as effective strategies for multiple cancer types. These peptides work by aiding the immune system in identifying and targeting cancer cells, offering the possibility of long-lasting therapeutic effects.

- **Viruses**

The COVID-19 pandemic prompted significant research into antiviral peptides, especially for creating peptide vaccines targeting SARS-CoV-2. Modern techniques such as immunoinformatics, epitope-based design, and molecular docking have accelerated the discovery and formulation of potential peptide vaccine candidates. Although there are no approved peptide vaccines for COVID-19 at this time, these endeavors have established a foundation for addressing future viral challenges.

Therapeutic peptides are rapidly evolving as potent tools for managing chronic illnesses and even potentially curing diseases. Their versatility, precision, and ability to target specific molecular pathways make them a promising avenue for improving healthcare outcomes. Whether it's optimizing glycemic control in diabetes, managing cardiovascular conditions, addressing gastrointestinal disorders, battling cancer, or countering viral infections, the development and application of therapeutic peptides continue to expand the horizons of modern medicine, offering hope for a healthier future.

PROACTIVE HEALTH: PEPTIDES IN PREVENTIVE CARE

Personalized peptide protocols have the potential to revolutionize preventive medicine and longevity. Peptides can have various effects on the body, including stimulating the release of human growth hormone, promoting muscle growth, and fighting off disease. Doctors can screen for individual patients' susceptibility to diseases using peptide arrays and tailor treatment plans accordingly. For example, peptide arrays can be used to identify people who are at risk of developing Alzheimer's disease. These patients may have a deficiency in certain peptides important for brain health. By identifying these patients early on, doctors can intervene with treatments that may help prevent or delay the onset of the disease.

Peptide arrays can aid in the early detection of individuals predisposed to cancer due to gene mutations linked to peptides involved in cell growth and division. Early identification allows for timely interventions that might inhibit or decelerate the proliferation of cancerous cells. Additionally, peptide arrays can pinpoint deficiencies in specific peptides associated with aging. By addressing these deficiencies, medical professionals can enhance individuals' longevity and overall health.

These are just a few examples of the potential applications of personalized peptide protocols in preventive medicine and longevity. As the technology continues developing, we will likely see more innovative ways to use peptides to improve human health.

PIONEERS IN PEPTIDE RESEARCH

Many biotech companies are working to develop peptide-based treatments for chronic illnesses. These pioneering initiatives are helping to bring us closer to a future where personalized medicine can prevent, treat, and cure disease. One such biotech company pioneering a groundbreaking peptide-based treatment for a chronic illness is Peptidream.

Peptidream is developing peptide-based therapies for neurodegenerative diseases. One of their lead candidates is AP101, a peptide designed to bind to and activate a protein called amyloid precursor protein (APP). APP is a key player in the development of Alzheimer's disease, and AP101 is thought to be able to prevent the formation of amyloid plaques, which are a hallmark of the disease. Peptidream is currently conducting clinical trials of AP101 in patients with Alzheimer's. The results of these trials have been promising, and Peptidream is hopeful that AP101 could be a breakthrough treatment for the disease.

The potential transformative impact of such initiatives is enormous. Alzheimer's disease affects millions of people worldwide. There is currently no cure for the disease, and existing treatments only offer modest relief. If AP101 or other peptide-based therapies are successful, they could revolutionize the treatment of Alzheimer's and improve the lives of millions of people.

The potential of peptide-based therapies is vast, and these pioneering initiatives are helping to pave the way for a new era of personalized medicine. Other companies that are pioneering peptide-based treatments for chronic illnesses include:

- **Alnylam Pharmaceuticals:** Alnylam is developing RNAi-based therapies that target disease-causing genes. One of their lead candidates is Patisiran, an RNAi therapy designed to treat hereditary transthyretin amyloidosis (ATTR), a rare and fatal genetic disorder.
- **ProQR Therapeutics:** ProQR is developing RNA editing therapies that can correct gene mutations that cause disease. One of their lead candidates is QR-110, an RNA editing therapy designed to treat cystic fibrosis.
- **Verily:** Verily is a subsidiary of Google, developing new ways to use technology to improve human health. One of their projects is called Project Baseline, a large-scale study collecting data on the health of thousands of people. This data is being used to develop new personalized treatments for disease.

A team of researchers from St Andrews University has developed peptides to help combat bacteria growing in biofilms. These peptides are called cationic antimicrobial peptides (CAMPs). CAMPs are short chains of amino acids that have a positive charge. This positive charge allows them to bind to the negatively charged surface of bacteria, thus killing them.

Biofilms are structured colonies of bacteria encased in a protective slimy matrix. They can form in diverse settings, including within the human body. A significant challenge with biofilms is their resistance to antibiotics and other antimicrobial treatments. In the healthcare sector, biofilm-associated infections are a major concern. They account for approximately 80% of all human infections, manifesting in situations like joint replacements, prosthetic devices, and catheter contaminations. Such infections are challenging to address and can result in severe outcomes, including sepsis and, in some cases, fatality.

Researchers evaluated the effectiveness of CAMPs against multiple bacterial strains, including antibiotic-resistant ones. The results indicated that CAMPs effectively combated all tested bacteria and could penetrate and eliminate bacteria within biofilms. These results point to the potential of CAMPs as a treatment for biofilm-associated infections. However, further studies are required to determine the safety and effectiveness of CAMPs in human applications.

As we stand on the cusp of what may be a transformative era in medicine, the horizon is gleaming with the potential of peptides. Their versatility and adaptability make them prime candidates for many therapeutic avenues. From tailored treatments addressing specific genetic dispositions to possibly rewriting the narrative of aging, peptides represent a beacon of hope. As technology and research evolve, it won't be surprising if the coming decades herald peptides as one of the cornerstones of modern therapeutic practice, fundamentally reshaping our approach to health, wellness, and longevity.

Now that you've seen the present and the future peptides let's take a moment to review the journey we've been on and the empowering knowledge you've acquired.

GLOSSARY OF PEPTIDES

This glossary provides an overview of selected peptides, synthetic analogs and peptide hormones discussed in this book. It is recommended that you speak to your healthcare provider should you wish to see an extensive list of peptides suitable for different purposes. Each entry provides the peptide's primary function and how its effectiveness is commonly evaluated:

ANTI-CANCER & IMMUNOTHERAPY

Name	Primary Use	Measure of Effectiveness
GnRH Analogs	Treatment of hormone-dependent cancers	Slowed progression of hormone-dependent cancers
PD-1 and CTLA-4-blocking Peptides	Cancer immunotherapy	Enhanced immune response against cancer cells
Somatostatin Analogs (e.g., Octreotide and Lanreotide)	Treating symptoms of certain tumors like carcinoid syndrome and acromegaly	Reduction in disease-related symptoms and hormone levels
Thymosin alpha-1 (Tα1)	Immune support	Improved immune function

ANTIMICROBIAL & PATHOGEN DEFENSE

Name	Primary Use	Measure of Effectiveness
Antimicrobial Peptides	Killing microorganisms and acne treatment	Ability to inhibit specific microorganisms
Bacitracin	Antimicrobial functions	Efficacy against specific bacteria
Cationic Antimicrobial Peptides (CAMPs)	Combat bacteria growing in biofilms	Effective elimination of bacteria within biofilms
Mersacidin	Bacterial peptide	Effectiveness against target organisms
Nisin	Bacterial peptide	Effectiveness against target organisms
Polymyxin B	Antimicrobial functions	Efficacy against specific bacteria

BONE HEALTH & REPAIR

Name	Primary Use	Measure of Effectiveness
Ostarine (MK-2866)	Promote bone health	Enhanced bone density

CARDIOVASCULAR & RENAL HEALTH

Name	Primary Use	Measure of Effectiveness
Atrial Natriuretic Peptide (ANP)	Protective roles for the heart and kidneys	Measurements related to heart growth and blood vessel dilation
Brain Natriuretic Peptide (BNP)	Protective roles for the heart and kidneys	Measurements related to heart growth and blood vessel dilation
C-type Natriuretic Peptide (CNP)	Protective roles for the heart and kidneys	Measurements related to heart growth and blood vessel dilation

COGNITIVE IMPROVEMENT & NEUROPROTECTION

Name	Primary Use	Measure of Effectiveness
AP101	Preventing the formation of amyloid plaques in Alzheimer's disease	Reduced formation of amyloid plaques
Cerebrolysin	Cognitive improvement and neuroprotection, particularly after strokes or brain injuries	Neurological recovery and cognitive function enhancement
DIHEXA	Potential treatment for Parkinson's and Alzheimer's	Improved brain function and clarity
FGL	Investigated for its potential role in enhancing synaptic plasticity and memory	Enhancement of learning and memory in experimental setups
Noopept	Prevention of cognitive impairment	Enhanced spatial memory
RG3	A ginsenoside with potential neuroprotective effects	Neuroprotective properties in experimental setups, particularly against oxidative stress and inflammation
Selank	Treatment for generalized anxiety disorder	Increased BDNF levels
Semax	Enhances learning and memory	Gene modulation and tempered immune response

HORMONAL REGULATION & METABOLISM

Name	Primary Use	Measure of Effectiveness
Adrenocorticotropic Hormone (ACTH)	Various bodily functions	Hormone level measurement and response in target organs
Antidiuretic Hormone (ADH)	Regulating water balance in the body	Maintaining balanced water levels
Glucagon	Elevating blood sugar levels when low	Measuring blood sugar levels
Growth Hormone-Releasing Peptide 6 (GHRP-6)	Stimulate growth hormone release	Increased growth hormone levels
Growth Hormone-Releasing Peptides (GHRPs)	Improve athletic performance, muscle recovery, and endurance	Athletic performance improvement and reduced recovery times
Ibutamoren (MK-677)	Stimulate the production of growth hormone	Growth hormone levels and side effects like fluid retention or increased appetite
Insulin	Regulating blood sugar levels	Monitoring blood sugar levels
Ipamorelin	Stimulate the production of growth hormone	Increased growth hormone levels
Oxytocin	Roles in social bonding and childbirth	Assessment of bonding behaviors and childbirth processes
Semaglutide	Weight loss and blood glucose regulation	Weight loss and improved blood glucose levels as the measure of effectiveness
Sermorelin	Stimulate the production of growth hormone	Increased growth hormone levels
Tesamorelin	Stimulate the production of growth hormone	Increased growth hormone levels

MUSCLE GROWTH & REPAIR

Name	Primary Use	Measure of Effectiveness
BPC-157	Healing injuries and accelerating recovery	Speed and quality of injury recovery and reduced inflammation
Cathelicidin-DM	Boosts cell growth and enhances wound healing	Wound healing rate and cell growth rate
CJC-1295	Stimulates the production of HGH and aids in cell regeneration and repair	HGH levels and rate of tissue repair
Creatine Peptides	Muscle building	Muscle mass and strength gains
Insulin-like Growth Factor 1 (IGF-1)	Aids tissue regeneration and muscle growth	Muscle growth and tissue regeneration rate
Mechano Growth Factor (MGF)	Stimulates damaged cells to divide and form new ones	Muscle repair and growth rate
Thymosin beta-4	Investigated for tissue repair, wound healing, and anti-inflammatory properties	Rate of tissue repair, wound healing, and reduction of inflammation

GLOSSARY OF PEPTIDES | 177

SKIN HEALTH & ANTI-AGING

Name	Primary Use	Measure of Effectiveness
Argireline	Wrinkle reduction in skincare	Reduction in wrinkle depth and appearance
Collagen Peptides	Improving skin and joint health	Reduced skin wrinkles, improved skin hydration, elasticity and joint flexibility
Epitalon	Investigated for potential anti-aging benefits	Potential extension of lifespan in experiments and possible telomere elongation
GHK-Cu	Tissue repair and anti-aging	Aids in wound healing and collagen production
Matrixyl	Collagen production	Reduced wrinkles and improved skin suppleness
Melanotan	Promoting skin tanning	Degree of skin tanning
Melanotan II	Investigated for promoting skin tanning and potential libido-enhancing effects	Darkening of the skin and potential improvement in sexual dysfunction

MISCELLANEOUS

Name	Primary Use	Measure of Effectiveness
AOD-9604	Investigated for weight loss and fat metabolism	Potential reduction in body fat
Beta-endorphin	Part of the Pro-opiomelanocortin (POMC) gene family	Measured by pain relief and mood elevation
Beta-lipotropin	Part of the Pro-opiomelanocortin (POMC) gene family	Depends on its specific functional role
Bremelanotide (PT-141)	Treatment of sexual dysfunction in men and women	Improved sexual arousal and reduced sexual dysfunction
Calcitonin Gene-related Peptide (CGRP)	Influence on energy levels and appetite	Evaluation of energy, appetite, and migraine occurrence
Chromofungin (CHR: CHGA47-66)	Initiates the release of other bioactive peptides and reduces inflammation	Reduced inflammation and associated symptoms
Liraglutide	Weight loss and blood glucose regulation	Weight reduction and stabilized blood glucose levels
Melanocyte-stimulating Hormone (MSH)	Various bodily functions	Evaluation based on intended functions such as skin pigmentation
PEDF	Inflammation and blood vessel formation reduction	Enhanced stem cell growth
PEDF-derived Short Peptide (PDSP)	Stem cell growth	Corneal cell regeneration
Secretin	Neutralizing stomach acid	Level of bicarbonate release and stomach acid neutralization

This list is by no means exhaustive, but it offers a glimpse into the diverse world of peptides and their applications. Whether naturally occurring or synthetically crafted, peptides and their analogs remain at the forefront of various therapeutic and cosmetic innovations.

CONCLUSION

Throughout this book, we have journeyed through the intricate world of peptides and their remarkable potential for health optimization. From understanding the basics of what peptides are to uncovering the myriad of ways they can influence our health–be it through enhanced physical performance, improved cognitive function, or faster healing–it's evident that peptides have a transformative role in contemporary medicine.

The exploration has underscored that peptide therapy is safe and effective when administered under a healthcare expert's careful guidance. This approach to treatment offers renewed hope and has the potential to enhance the health and well-being of many individuals significantly.

Now, with a foundation in peptide knowledge and the ability to create a personalized peptide protocol, the path forward beckons you to dig deeper. The universe of peptides is vast, and the wealth of resources available–both online and traditional–will be invaluable companions in your continued exploration. It's always advisable to consult with healthcare professionals as you delve further, ensuring that any decisions are informed and best suited for your unique needs.

In an age where science and medicine are progressing at a pace previously unimagined, peptides stand at the forefront of this evolution. It is exciting to be involved and informed about these biological marvels. As you navigate the vast ocean of peptide research, always remember the importance of staying abreast of the latest advancements. Knowledge is power, and with power comes the ability to make the best choices for optimal health.

Reflecting on this comprehensive guide, it's clear that peptides are not just a fleeting topic in the annals of medicine; they represent a promising frontier. This book is intended to educate and inspire, fueling curiosity and passion for the many possibilities of peptides.

As our exploration concludes, your thoughts and feedback are invaluable. If *The Secret Power of Therapeutic*

Peptides: How to Use Innovative Science to Build Muscle Faster, Heal from Injury, and Boost Your Focus has enlightened and enriched your understanding of peptides, please consider taking a few moments to leave a review. Your insights and reflections can help future readers embark on their own journey through the world of peptides, and your support is instrumental in sharing the importance of this groundbreaking field. Remember, every perspective counts, and yours could be the catalyst for another's discovery and passion. Once again, thank you for accompanying us on this journey, and here's to the transformative power of peptides.

REFERENCES

Aging, cell division | Learn science at Scitable. (n.d.). Nature. https://www.nature.com/scitable/topicpage/aging-and-cell-division-14230076/#:

Al-Atif, H. (2022). Collagen supplements for aging and wrinkles: A paradigm shift in the Field of dermatology and cosmetics. Dermatology Practical & Conceptual, e2022018. https://doi.org/10.5826/dpc.1201a18

Alina Petre, A. P. (2020, December 3). Peptides for bodybuilding: Do they work, and are they safe? Healthline. https://www.healthline.com/nutrition/peptides-for-bodybuilding#what-they-are

Amino acids, evolution | Learn science at Scitable. (n.d.). https://www.nature.com/scitable/topicpage/an-evolutionary-perspective-on-amino-acids-14568445/

Asua, D., Bougamra, G., Calleja-Felipe, M., Morales, M., & Knafo, S. (2018). Peptides acting as cognitive enhancers. Neuroscience, 370, 81-87. https://doi.org/10.1016/j.neuroscience.2017.10.002

Bajnath, A. (2023, January 28). Anti-aging with peptides: The future of longevity medicine? The Institute for Human Optimization. https://ifho.org/anti-aging-with-peptides-the-future-of-longevity-medicine/

Barrett, G. C., & Elmore, D. T. (1998). Amino acids and peptides. Cambridge University Press.

Ben Greenfield: Peptides for Biohacking. (2023, June 5). LinkedIn. https://www.linkedin.com/pulse/ben-greenfield-peptides-biohacking-getblokes

Biochemistry, peptide - StatPearls - NCBI bookshelf. (2022, August 29). National Center for Biotechnology Information. https://www.ncbi.nlm.nih.gov/books/NBK562260/#:

Boulder Longevity Institute. (2023, April 18). How to use peptides to shorten recovery time between workouts. https://boulderlongevity. com/use-peptides-for-faster-recovery-time-between-workouts/#:

Brain health. (2020, June 3). World Health Organization (WHO). https://www.who.int/health-topics/brain-health#tab=tab_1

Brinkmann, J., Voskuhl, J., & Jonkheijm, P. (2018). Peptide and protein printing for tissue regeneration and repair. Peptides and Proteins as Biomaterials for Tissue Regeneration and Repair, 229-243. https://doi.org/10.1016/b978-0-08-100803-4.00009-7

Campisi, J., & Yaswen, P. (2009). Aging and cancer cell biology, 2009. Aging Cell, 8(3), 221-225. https://doi.org/10.1111/j.1474-9726.2009.00475.x

Carlson, M. C., Moored, K., Rebok, G. W., & Eaton, W. W. (2019). Public mental health and the brain across the life span. Public Mental Health, 257-284. https://doi.org/10.1093/oso/9780190916602.003.0011

Cognitive function. (n.d.). Advitam | Peptide Therapy & Hormone Replacement Therapy located in Midtown Manhattan, New York, NY. https://www.myadvitam.com/content/cognitive-function#:

Devaraj, E., Anbalagan, M., Ileng Kumaran, R., & Bhaskaran, N. (2021). Immunity, stem cells, and aging. Stem Cells and Aging, 89-101. https://doi.org/10.1016/b978-0-12-820071-1.00006-2

Fried, S. D., Fujishima, K., Makarov, M., Cherepashuk, I., & Hlouchova, K. (2022). Peptides before and during the nucleotide world: An origins story emphasizing cooperation between proteins and nucleic acids. Journal of The Royal Society Interface, 19(187). https://doi.org/10.1098/rsif.2021.0641

Gomes, A., Teixeira, C., Ferraz, R., Prudêncio, C., & Gomes, P. (2017). Wound-healing peptides for treatment of chronic diabetic foot ulcers and other infected skin injuries. Molecules, 22(10), 1743. https://doi.org/10.3390/molecules22101743

Greenfield, B. (2022, June 6). BPC 157: How to use it to healing your body like wolverine. Ben Greenfield Life - Health, Diet, Fitness, Family & Faith. https://bengreenfieldlife.com/article/supplements-articles/how-to-use-bpc-157/

Holzer, P., & Farzi, A. (2020). Gut hormones and neuropeptides as

REFERENCES | 185

mediators of microbiome-brain communication. The Oxford Handbook of the Microbiome-Gut-Brain Axis. https://doi.org/10.1093/oxfordhb/9780190931544.013.1

Holzer, P., Reichmann, F., & Farzi, A. (2012). Neuropeptide Y, peptide YY and pancreatic polypeptide in the gut–brain axis. Neuropeptides, 46(6), 261-274. https://doi.org/10.1016/j.npep.2012.08.005

How to improve brain health and cognition with peptides. (2020, 9). Dr. Neil Paulvin – Health, Wellness & Integrative Medicine in New York. https://doctorpaulvin.com/blog/how-to-improve-brain-health-and-cognition-with-peptides/

How to improve brain health and cognition with peptides. (2020, 9). Dr. Neil Paulvin – Health, Wellness & Integrative Medicine in New York. https://doctorpaulvin.com/blog/how-to-improve-brain-health-and-cognition-with-peptides/

Janice C. Froehlich. (1995). The role of opioid peptides in environmentally-induced analgesia. Pharmacology of Opioid Peptides, 359-410. https://doi.org/10.1201/9781482264487-24

Kara Rodgers. (n.d.). What is the difference between a peptide and a protein? Encyclopedia Britannica. https://www.britannica.com/story/what-is-the-difference-between-a-peptide-and-a-protein#:

La Manna, S., Di Natale, C., Florio, D., & Marasco, D. (2018). Peptides as therapeutic agents for inflammatory-related diseases. International Journal of Molecular Sciences, 19(9), 2714. https://doi.org/10.3390/ijms19092714

Lau, J. L., & Dunn, M. K. (2018). Therapeutic peptides: Historical perspectives, current development trends, and future directions. Bioorganic & Medicinal Chemistry, 26(10), 2700-2707. https://doi.org/10.1016/j.bmc.2017.06.052

Leonard, J. (n.d.). Peptides: What are they, uses, and side effects. Medical and health information. https://www.medicalnewstoday.com/articles/326701#about

Martínez-Villaluenga, C., & Hernández-Ledesma, B. (2020). Peptides for health benefits 2019. International Journal of Molecular Sciences, 21(7), 2543. https://doi.org/10.3390/ijms21072543

A new peptide may hold potential as an Alzheimer's treatment. (n.d.). MIT News | Massachusetts Institute of Technology. https://news.mit.edu/2023/new-peptide-may-hold-potential-alzheimers-treatment-0413

OA prevalence and burden. (2022, October 12). Osteoarthritis Action Alliance. https://oaaction.unc.edu/oa-module/oa-prevalence-and-burden/#:

Peptide therapeutics market size, report 2032. (n.d.). Precedence Research - Market Research Reports & Consulting Firm. https://www.precedenceresearch.com/peptide-therapeutics-market#:

Peptides for bodybuilding: Does it actually work? (2021, May 15). Transparent Labs. https://www.transparentlabs.com/blogs/all/peptides-for-bodybuilding-how-growth-hormone-secretagogues-work#:

Peptides For Muscle Growth: Do They Work? (2023, July 31). Healthnews. https://healthnews.com/nutrition/vitamins-and-supplements/peptides-for-muscle-growth-do-they-work/

Pieds. (2023, January 10). The Alcohol and Drug Foundation - Alcohol and Drug Foundation. https://adf.org.au/drug-facts/pieds/

Pountos, I., Panteli, M., Lampropoulos, A., Jones, E., Calori, G. M., & Giannoudis, P. V. (2016). The role of peptides in bone healing and regeneration: A systematic review. BMC Medicine, 14(1). https://doi.org/10.1186/s12916-016-0646-y

Researchers develop new measure of brain health. (2023, September 11). ScienceDaily. https://www.sciencedaily.com/releases/2021/06/210601100640.htm

Salame, B., & D.C. (2021, September 13). To gain an edge, athletes, biohackers, and celebrities are increasingly turning to these molecules. Medium. https://medium.com/in-fitness-and-in-health/for-an-edge-athletes-biohackers-and-celebrities-are-increasingly-turning-to-these-molecules-8ecbb45935e3

Singh, A. (2022, April 21). Classes and benefits of peptides. Conduct Science. https://conductscience.com/classes-and-benefits-of-peptides/

Snyder, S. H. (1980). Brain peptides as neurotransmitters. Science, 209(4460), 976-983. https://doi.org/10.1126/science.6157191

Tech, A. C. (2022, February 20). What are peptides actually used for? Advanced ChemTech - Providing Various Fine Chemicals in Louisville, KY. https://www.advancedchemtech.com/what-are-peptides-actually-used-for/#:

Unlocking your brain's potential: The top neurotropics and peptides for cognitive enhancement. (n.d.). LinkedIn. https://www.linkedin.com/pulse/unlocking-your-brains-potential-top-neurotropics-peptides-jimmy-brady/

Unsafe, untested and unregulated. (2013, February 7). The Sydney Morning Herald. https://www.smh.com.au/politics/federal/unsafe-untested-and-unregulated-20130207-2e1cc.html

Wang, G., Chen, Z., Tian, P., Han, Q., Zhang, J., Zhang, A., & Song, Y. (2022). Wound healing mechanism of antimicrobial peptide cathelicidin-DM. Frontiers in Bioengineering and Biotechnology, 10. https://doi.org/10.3389/fbioe.2022.977159

Wang, L., Wang, N., Zhang, W., Cheng, X., Yan, Z., Shao, G., Wang, X., Wang, R., & Fu, C. (2022). Therapeutic peptides: Current applications and future directions. Signal Transduction and Targeted Therapy, 7(1). https://doi.org/10.1038/s41392-022-00904-4

Wang, Y., Pan, Y., & Li, H. (2020). What is brain health and why is it important? BMJ, m3683. https://doi.org/10.1136/bmj.m3683

What are peptides. (2023, April 25). Zealand Pharma A/S. https://www.zealandpharma.com/rd/what-are-peptides/#:

What are peptides? What to know about this anti-aging ingredient. (2022, January 4). Health. https://www.health.com/beauty/skincare/what-are-peptides#:

Williams, A. (2023, March 20). The Wall Street Journal - The next fountain-of-Youth craze? Peptide injections. Upgrade Labs | Human Upgrade™ Center. https://upgradelabs.com/the-wall-street-journal-the-next-fountain-of-youth-craze-peptide-injections/

Wizard. (2022, February 1). Peptide therapy and sexual enhancement. Medical Spa: Your Skin and Health is our (PRIORITY). https://

ajeless.com/2022/02/01/peptide-therapy-and-sexual-enhancement/#:

Zaky, A. A., Simal-Gandara, J., Eun, J., Shim, J., & Abd El-Aty, A. M. (2022). Bioactivities, applications, safety, and health benefits of Bioactive peptides from food and by-products: A review. Frontiers in Nutrition, 8. https://doi.org/10.3389/fnut.2021.815640

Zane, D., Feldman, P. L., Sawyer, T., Sobol, Z., & Hawes, J. (2020). Development and regulatory challenges for peptide therapeutics. International Journal of Toxicology, 40(2), 108-124. https://doi.org/10.1177/1091581820977846

Made in the USA
Middletown, DE
26 October 2023